Rare Earth Elements - Emerging Advances, Technology Utilization, and Resource Procurement
http://dx.doi.org/10.5772/intechopen.102266
Edited by Michael T. Aide

Contributors
Moses Radebe, Manny Mathuthu, Iwona Cieślik, Monika Duchna, Damien Krebs, Domenic Furfaro, Damián Reyes-Jáquez, Efren Delgado, Hiram Medrano-Roldán, Francisco Adrián De la Torre Martínez, María Dolores Josefina Rodriguez Rosales, Javier López Miranda, Claudio Pettinari, Andrei Drozdov, Yuriy Belousov, Michael T. Thomas Aide, Jiangang Jiao, Refaey El-Wardany

First published in London, United Kingdom, 2023 by IntechOpen
IntechOpen is the global imprint of INTECHOPEN LIMITED, registered in England and Wales, registration number: 11086078, 5 Princes Gate Court, London, SW7 2QJ, United Kingdom

British Library Cataloguing-in-Publication Data
A catalogue record for this book is available from the British Library

Additional hard and PDF copies can be obtained from orders@intechopen.com

Rare Earth Elements - Emerging Advances, Technology Utilization, and Resource Procurement
Edited by Michael T. Aide
p. cm.
Print ISBN 978-1-83768-074-0
Online ISBN 978-1-83768-075-7
eBook (PDF) ISBN 978-1-83768-076-4

Meet the editor

Dr. Michael Aide received his Ph.D. in Soil Chemistry from Mississippi State University, USA, in 1982, and a bachelor's degree in Chemistry and Mathematics from the University of Wisconsin-Madison, USA. He has been an educator and agronomic-soil researcher at Southeast Missouri State University, USA, since 1982. His research interests include the growth and development of rice in integrated systems involving soil fertility, water management, and mitigation of arsenic uptake. Rice research has permitted his travel to southeastern Asia, Central America, and Egypt. Dr. Aide is a certified soil scientist who has investigated rare earth elements in soils, attempting to utilize them to indicate parent material uniformity and fate/transport. His professional affiliations include the Soil Science Society of America and the Soil and Water Conservation Society.

Contents

Preface

Rare Earth Elements - Emerging Advances, Technology Utilization, and Resource Procurement presents a compelling array of research detailing aspects of the technology and utilization of rare earth elements and similar elements. Chapters in the section on "Emerging Advances and Technology" discuss coordination compounds involving lanthanides to develop luminescent turn-off sensors, gold mineralization in Egypt, concentrated hydrochloric acid leaching of steenstrupine to obviate silica gel formation, luminescent materials with turn on and radiometric sensory response based on coordination compounds, prediction of solubility and miscibility parameters of the bismuth–arsenic complex and amorphous mineral compounds using molecular dynamic simulation, rare earth elements in new advanced engineering applications, and a radon emanation measurement system for South African communities. The section on "Utilization and Resource Procurement" presents an evaluation of rare earth element mine sites for environmental impact.

Each chapter provides a unique perspective on the uses or technology associated with rare earth elements. Collectively, they showcase the diversity of research involving these elements.

Dr. Michael T. Aide
Professor of Agriculture,
Southeast Missouri State University,
Missouri, USA

Section 1

Emerging Advances and Technology

Chapter 1

Coordination Compounds of Lanthanides as Materials for Luminescent Turn Off Sensors

Claudio Pettinari, Andrei Drozdov and Yuriy Belousov

Abstract

This review aims at describing the possible use of lanthanide coordination compounds as materials for luminescent sensors now more necessary due to the continuous requirements from the society of electroluminescent and lighting devices, for example analytical sensors and imaging instruments. This is the first part of a work describing the photophysical foundations of the luminescence of complex compounds of lanthanides in the context of design materials with a sensory response, and also considers in detail materials with the most common type of response - *turn off* sensors.

Keywords: lanthanides, coordination compounds, luminescence, sensors, sensory response mechanism, turn off sensors

1. Introduction

The use of coordination compounds, organometallic and inorganic species containing lanthanides (Ln) as luminescent materials is mainly due to electronic transitions inside f-shells [1–6]. The design of luminescent sensors based on lanthanides is currently catching up and the number of papers on emitting materials, in particular Organic Light-Emitting Diode (OLED), is increasing. Several lanthanide complexes and Metal Organic Frameworks (MOFs) were described as useful for ratiometric fluorescent sensing [2] or potentially employable as functional materials [5]. These compounds are often characterized by switchable and tunable properties allowing fine-tuned optical features and sensitive responses to small molecules and ions. Several reviews [7–12] have reported lanthanide-containing luminescent sensor materials, which can be used for detecting anions [9] or low molecular weight analytes [8], and also for detecting cations [12]. The rapid growth in the number of publications requires a systematization that could help in the choice of the right compounds for new devices. Here we report a general overview of the principles governing the lanthanide coordination compounds luminescence and the most important examples of Ln compounds for sensing both ions and low molecular weight compounds. We have decided to report here a classification of the species based on the type of *sensory signal being recorded*, in detail "turn off", "turn on" and "ratiometric" sensors.

2. Lanthanide complexes luminescence principles

To understand how to control a sensory response by using a lanthanide complex as sensor material, it is necessary to consider the main processes occurring when Ln^{3+} compounds exhibit luminescence. In the absence of organic ligand environments, absorption and emission are due to f-f transitions in the electron shells of lanthanides which, at least theoretically, are possible for all ions on going from Ce^{3+} $(4f^{15}d^{1})$ to Yb^{3+} $(4f^{13})$. At the same time, the structure of the levels first becomes progressively more complicated on going from Ce^{3+} to Eu^{3+}, and then gradually becomes simpler on going from Tb^{3+} to Yb^{3+} ion. It is relevant that symmetric ions with respect to Gd^{3+}, located in the center of the Ln series, have some similarity both in the structure of electronic levels and in luminescence features, such as the lifetime of the excited state (**Figure 1**).

The largest energy gap between the excited and ground levels is observed for the Gd^{3+} ion and corresponds to UV emission. Among those emitting in the visible region (**Table 1**), the most efficient phosphors are Tb^{3+} and Eu^{3+} ions; Sm^{3+} and Dy^{3+} ions are

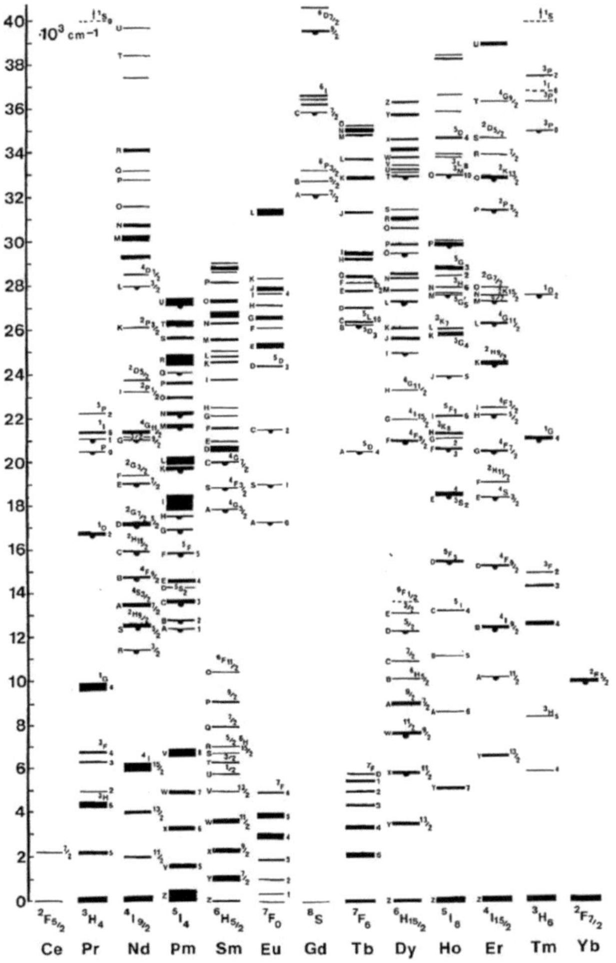

Figure 1.
Free ion energy levels of the trivalent lanthanide ions from Ce^{3+} $(4f^{1})$ to Yb^{3+} $(4f^{13})$. Levels are labeled by term symbols or, for some higher levels, capital letters. Reprinted from Dieke and Crosswhite, 39 copyright 1963, with permission from the Optical Society of America [13].

Ion, electronic configuration	Excited state	Emission ground state	λ_{EM}, nm	Emission color	τ_{H2O}, μs [ref]	τ_{D2O}, μs [ref]
Ce [Xe]4f^1	^2D	$^2F_{7/2}$ $^2F_{5/2}$	351 365	UV		
Pr [Xe]4f^2	$^3P_0$ (20500) $^1D_2$	^3H_j, j = 4,5,6 $^3F_4$	488, 528, 6,111,050	Red NIR		
Nd[Xe]4f^3	$^4F_{3/2}$ (11460)	^4I_j, j = $^9/_2$–$^{13}/_2$ j = $^{15}/_2$–$^9/_2$	900, 1060, 1350	NIR	0.031 [14] 0.031 [15]	0.14 [15]
Sm [Xe]4f^5	$^4G_{5/2}$ (17900) $^4G_{5/2}$, $^4G_{5/2}$	^6H_j, j = $^5/_2$–$^{13}/_2$ ^6F_j, j = $^1/_2$–$^9/_2$ $^6H_{13/2}$	560, 595, 640, 700, 775 870, 887, 926, 1010 1150 877	Orange, NIR	2.69 [14] 2.7 [15] 10 $SmCl_3 . 6H_2O$	60 [15]
Eu [Xe]4f^6	$^5D_0$ (17300) $^5D_1$ $^5D_2$	^7F_j, j = 0–6 ^7F_j, j = 0–4 ^7F_j, j = 0–4	580, 590, 615, 650, 720, 750, 820 525, 535, 557, 585, 625 462, 470, 487, 510, 535	Red	112.4 [14] 108 [15]	4020 [14] 4100 [15]
Gd [Xe]4f^7	$^6P_{7/2}$ (32200)	$^8S_{7/2}$	315	UV	1480 [14]	
Tb [Xe]4f^8	$^5D_4$ (20500)	^7F_j, j = 6–0	490, 545, 580, 620, 650, 660, 675	Green	442 [14] 467 [15]	3800 [15]
Dy [Xe] 4f^9	$^4F_{9/2}$ (21100) $^4I_{15/2}$	^6H_j, j = $^{15}/_2$–$^9/_2$ ^6H_j, j = $^{15}/_2$–$^9/_2$	475, 570, 660, 750 455, 540, 615, 695	Yellow- green	2.49 [14] 2.6 [15]	42 [15]
Ho [Xe]4f^{10}	$^5F_5$ $^5I_6$ $^5F_4$ (18500)	$^5I_7$ $^5I_8$ ^5I_j, j = 8,7,6	965 1180 540, 750, 1015	NIR NIR Green +NIR		
Er [Xe]4f^{11}	$^4S_{3/2}$ $^4F_{9/2}$ ^4I_j, j = 9,11,13 (11750)	$^4I_{15/2}$ $^4I_{15/2}$ $^4I_{15/2}$	550 670 850, 980, 1560	Green Red NIR		
Tm [Xe]4f^{12}	$^3H_4$ (12400) $^1G_4$	$^3F_4$ $^3H_6$ $^3H_6$	1470 808 480	NIR NIR Blue		
Yb [Xe]4f^{13}	$^2F_{5/2}$ (10250)	$2F_{7/2}$	980	NIR	0.17 [15]	3.95 [15]

[a] *λ_{EM} – emission wavelenght, τ_{H2O} and τ_{D2O} – observed excitation state lifetime of Ln^{3+} solvated with H_2O or D_2O respectively.*

Table 1.
Ions emitting in the visible region.[a]

somewhat lower emitters. Among the IR-emitting ions, the most effective is Yb^{3+}, as well as Nd^{3+} and Er^{3+}. Ce^{3+} luminescence is possible due to d-f electronic transitions. Other ions (Pr^{3+}, Pm^{3+}, Ho^{3+}, Tm^{3+}) can exhibit emission in the visible and/or IR regions, however, as a rule, their intensity is low and difficult to detect with a common equipment. These considerations explain why europium and terbium compounds are used in most of the productions of lanthanide-based luminescent sensors.

With the exception of the d-f transitions for Ce^{3+} and Tb^{3+}, the electronic transitions of lanthanides are characterized by low extinction coefficients (**Table 2**). The extinction coefficients are significantly lower not only with respect to those found for organic molecules but also with respect to d-cations which cause a low luminescence brightness of most inorganic REE compounds. In 1942, Weissman proposed a solution to this problem [16]. In chelate complexes containing organic ligands characterized by conjugated aromatic fragments, absorption and emission are spatially separated due to the "antenna" effect: the organic molecule effectively absorbs radiation, sequentially passing to singlet and triplet excited states, and then it can transfer

Ion	Transition	Λ	ε, M^{-1} cm^{-1}
Ce^{3+}	$4f^1 \rightarrow 5d$	253	755
	$^2F_{5/2} \rightarrow {}^2F_{7/2}$	~5000	n/d
Pr^{3+}	$^3H_4 \rightarrow {}^3P_2$	444.0	10.1
	$^3H_4 \rightarrow {}^3P_1$	468.8	4.4
	$^3H_4 \rightarrow {}^3P_0$	482.2	4.1
	$^3H_4 \rightarrow {}^1D_2$	589.0	4.95
Nd^{3+}	$^4I_{9/2} \rightarrow {}^4D_{3/2}$	354.0	2.30
	$^4I_{9/2} \rightarrow {}^4G_{9/2}$	512.0	1.89
	$^4I_{9/2} \rightarrow {}^4G_{7/2}$	522.0	3.74
	$^4I_{9/2} \rightarrow {}^4G_{5/2}, {}^2G_{7/2}$	575.2	6.00
	$^4I_{9/2} \rightarrow {}^4S_{3/2}$	731.5	3.80
	$^4I_{9/2} \rightarrow {}^4F_{7/2}$	740.0	6.22
	$^4I_{9/2} \rightarrow {}^4F_{5/2}$	794.0	8.10
	$^4I_{9/2} \rightarrow {}^2H_{9/2}$	801.0	5.48
	$^4I_{9/2} \rightarrow {}^4F_{3/2}$	865.0	3.10
Sm^{3+}	$^6H_{5/2} \rightarrow {}^6P_{3/2}$	401.5	1.25
	$^6H_{5/2} \rightarrow {}^6F_{9/2}$	1089.0	2.19
	$^6H_{5/2} \rightarrow {}^6F_{7/2}$	1250	2.0
Eu^{3+}	$^7F_0 \rightarrow {}^5F_4$	298.0	1.04
	$^7F_0 \rightarrow {}^5H_6$	317.5	0.98
	$^7F_0 \rightarrow {}^5L_6$	396.0	2.90
Gd^{3+}	$^8S_{7/2} \rightarrow {}^6I_{13/2}$	272.7	3.16
	$^8S_{7/2} \rightarrow {}^6I_{11/2}$	273.0	1.12
	$^8S_{7/2} \rightarrow {}^6I_{9/2}, {}^6I_{7/2}$	275.6	1.90
Tb^{3+}	$4f_8 \rightarrow 4f_7{\cdot}5d$	219	374
	$^7F_6 \rightarrow {}^5G_5$	377	n/d
	$^7F_6 \rightarrow {}^5D_3$	385	n/d
	$^7F_6 \rightarrow {}^5D_4$	492	n/d

Ion	Transition	Λ	ε, M^{-1} cm^{-1}
Dy^{3+}	$^6H_{13/2} \rightarrow {}^6P_{3/2}$	325.0	1.88
	$^6H_{13/2} \rightarrow {}^6P_{7/2}$	350.5	2.60
	$^6H_{13/2} \rightarrow {}^4M_{19/2}$	365.0	2.14
	$^6H_{13/2} \rightarrow {}^6F_{5/2}$	807.0	1.84
	$^6H_{13/2} \rightarrow {}^6F_{7/2}$	908.5	2.40
	$^6H_{13/2} \rightarrow {}^6F_{9/2}, {}^6H_{7/2}$	1102.0	1.80
	$^6H_{13/2} \rightarrow {}^6F_{11/2}, {}^6H_{9/2}$	1300.0	1.07
Ho^{3+} Ce^{3+} Tm^{3+} Yb^{3+}	$^5I_8 \rightarrow {}^3F_4, {}^5D_4$	241.0	3.18
	$^5I_8 \rightarrow {}^3H_4$	278.0	2.21
	$^5I_8 \rightarrow {}^3D_4$	287.5	3.10
	$^5I_8 \rightarrow {}^3H_6$	361.5	2.10
	$^5I_8 \rightarrow {}^3G_5$	416.1	2.52
	$^5I_8 \rightarrow {}^3G_6$	452.0	3.70
	$^5I_8 \rightarrow {}^5F_3$	485.5	1.75
	$^5I_8 \rightarrow {}^5F_4$	536.5	4.55
	$^5I_8 \rightarrow {}^5F_5$	641.0	3.04
	$^4I_{15/2} \rightarrow {}^4D_{7/2}$	255.0	6.66
	$^4I_{15/2} \rightarrow {}^4G_{11/2}$	379.2	6.90
	$^4I_{15/2} \rightarrow {}^4F_{7/2}$	487.0	2.03
	$^4I_{15/2} \rightarrow {}^2H_{11/2}$	523.0	3.28
	$^4I_{15/2} \rightarrow {}^4F_{9/2}$	652.5	2.04
	$^3H_6 \rightarrow {}^3P_2$	262.0	1.05
	$^3H_6 \rightarrow {}^3D_2$	360.0	0.93
	$^3H_6 \rightarrow {}^3F_3$	683.0	2.36
	$^3H_6 \rightarrow {}^3H_4$	787.4	1.00
	$^3H_6 \rightarrow {}^3H_5$	1230.0	1.00
	$^2F_{7/2} \rightarrow {}^2F_{5/2}$	973	2.1

Table 2.
Absorption coefficients of Ln^{3+} ions corresponding to the most significant transitions [15].

excitation energy to the Ln^{3+} ion, which in turn emits a typical narrow-band lanthanide-centered emission [17]. In **Figure 2**, a scheme showing the emission mechanism of the luminescence of lanthanide complexes is reported.

It is worth considering the Jabłoński diagram in **Figure 3** to better understand the principles governing luminescence for sensory materials.

Processes occurring during antenna sensitization of lanthanide cations are schematically presented in the Jabłoński diagram (**Figure 3**). Light absorption (A) occurs mainly due to the chromophore groups of the ligand, and the extinction coefficient of the ligand can exceed that of the lanthanide ion by several orders of magnitude [18]. The coordinated ligand is excited to one of the singlet excited states ***S****, the transition

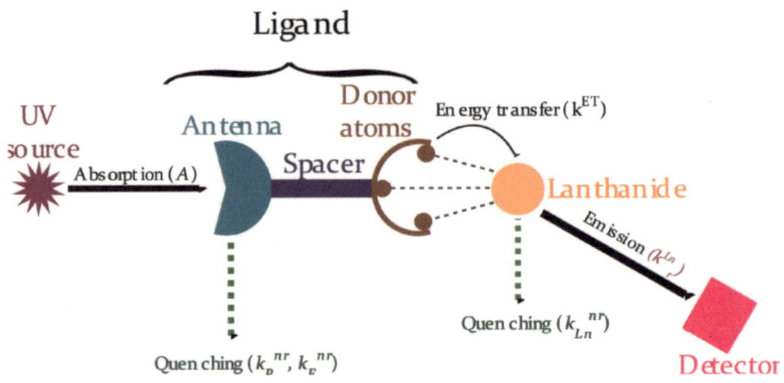

Figure 2.
The emission process for a Ln^{3+} luminescent complex. Antenna, spacer, coordination site, and central ion are evidenced.

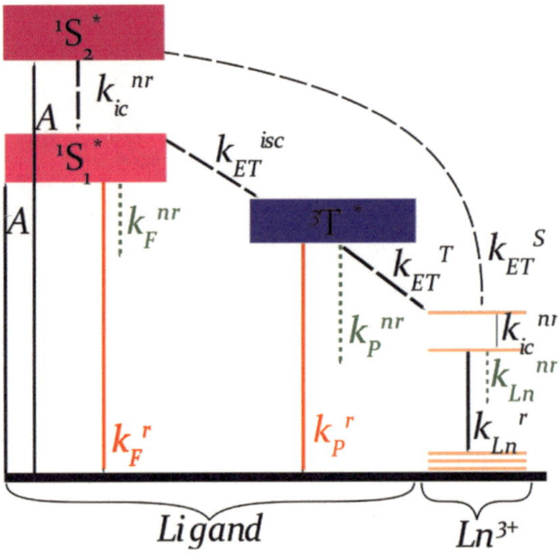

Figure 3.
Simplified Jabłoński diagram for luminescent Ln^{3+} complexes. The processes of absorption (A), radiative relaxation (abbreviated as r) – fluorescence (F), phosphorescence (P), and emission of lanthanides (Ln) are shown by straight lines; nonradiative relaxation processes (nr) are indicated by dotted lines; energy transfer (ET), internal conversion (ic), and intercombination conversion (isc) are indicated by dashed-dotted lines. The reverse energy transfer processes are not shown for clarity.

to the most stable state being defined as internal conversion (k_{ic}^{nr}). At this stage, the ligand can either emit a quantum of light (ligand fluorescence, rate constant k_F^r), or non-radiatively relax (k_F^{nr}), or its excited state S^* energy can be transferred to the triplet excited state $^3T^*$ by inter-combination conversion (k_{isc}). This triplet state is relatively long-lived due to the forbiddance of singlet-triplet transitions. The luminescent relaxation of the triplet excited state is called phosphorescence (k_P^r) and is well observed when the energy of $^3T^*$ is lower than the resonant level of lanthanide. This is usually the most common phenomenon for Gd^{3+} complexes [19]. The triplet excited state life is longer than that of the singlet one. Various relaxation pathways are possible for it: ligand phosphorescence (k_P^r), nonradiative relaxation (k_P^{nr}) due to energy transfer into lattice vibrations (multiphonon relaxation), and energy transfer to the lanthanide ion (k^T_{ET}). In addition to this pathway, in rare cases, direct excitation of the lanthanide ion through the singlet state of the ligand ($k_{ET}{}^{S}$) is also recorded. The transitions inside the f-shells of lanthanides are quite fast and are accompanied by the luminescence of the latters ($k_{Ln}{}^{r}$). Nonradiative relaxation of the excited states of lanthanides due to phonon lattice vibrations is also possible, which corresponds to the constant $k_{Ln}{}^{nr}$.

Thus, the luminescence efficiency of monometallic lanthanide complexes depends on the rate of many processes, and minor changes in the system can dramatically affect it. To date, some rules have been developed that are necessary for the production of efficient luminescent materials based on REE compounds, primarily Eu^{3+} and Tb^{3+}, but also Sm^{3+}, Dy^{3+}, IR-emitting Nd^{3+}, Yb^{3+}, and some other REE ions [17, 19–21].

3. Classification of sensory materials

There are several ways to classify sensory materials:

i. according to the type of response;

ii. according to the structural features;

iii. according to the Ln^{3+} ions employed;

iv. according to the mechanisms leading to the occurrence of the response.

In this manuscript we will consider examples of various materials, classifying them according to the type of recorded response, and evidencing also other relevant features.

From a phenomenological point of view, the most rational classification is on the basis of the type of *sensory signal being recorded*, the most common being "turn on", "turn off" and "ratiometric". These types of sensory response are unevenly represented in the literature (**Figure 4**). The most common type of response, i.e. the "turn off", is characterized by the least selectivity and unfortunately is not optimal for practical use.

It makes sense to distinguish the materials according to their structural features and typology. For example, it is useful to differentiate between soluble mononuclear

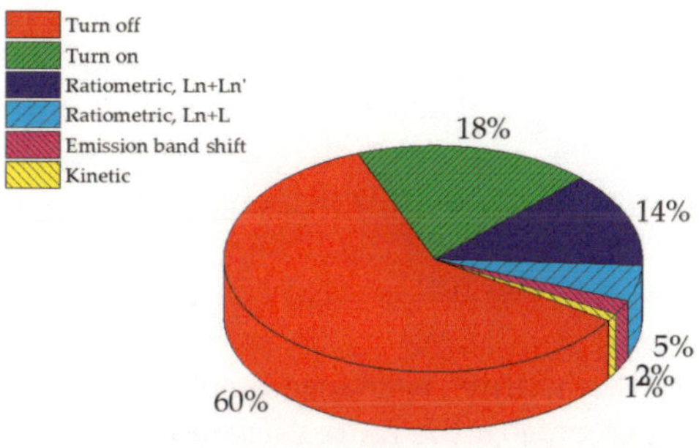

Figure 4.
Percentage of different types of sensory response in lanthanide-based luminescent sensors.

complexes, coordination polymers (CP's) and porous Metal–Organic Frameworks (MOF's) [22], or immobilized species on an inorganic support surface. Each of these materials has both advantages and disadvantages, and it is impossible to unambiguously define them as optimal without specifying the conditions for which they are optimal. For example, molecular complexes quickly react with an analyte and are convenient objects for studying the response mechanism; however, their regeneration is complicated, and they contaminate the sample. Complexes immobilized on carriers react more slowly and their preparation is time-consuming. Synthesis of insoluble MOF's is often even more difficult, but they offer the advantages of high chemical stability and increased selectivity due to limited pore and channel permeability.

A third possible classification of materials is according to the type of luminescent centers used:

a. materials that contain cations that effectively give luminescence, for example, Tb^{3+} or Eu^{3+};

b. materials that contain less intense emitters, for example, Sm^{3+} or Dy^{3+};

c. species containing lanthanide ions often emitting in the IR region (Nd^{3+}, Er^{3+}, Yb^{3+});

d. materials where the luminescence is exclusively due to the ligand;

e. bi- and polymetallic systems with several luminescent centers.

The luminescence efficiency of Eu^{3+} and Tb^{3+} compounds usually turn out to be much higher, moreover, their emission bands lie in the visible region (**Table 1**), where the sensitivity of standard photomultipliers is high. As a result, most of the described sensor materials are built by using Eu^{3+} and Tb^{3+} compounds for "turn on", "turn off", and "ratiometric" sensors. The luminescence of IR-emitting ions Nd^{3+}, Er^{3+}, and Yb^{3+} are more sensitive to C–H bond vibrations, which makes them also promising to produce special sensor materials.

Finally, the fourth classification method allows one to distinguish between materials according to the mechanism of sensory response. The proposed classification is based on the effect that the analyte has on both the structure of the complex and on the processes reflected in the Jabłoński diagram.

The most obvious is the mechanism of action defined "**internal filter**" **effect** (**IFE**), in which the absorption spectrum of the analyte overlaps with the excitation or luminescence band of the sensor material. This mechanism (**Figure 5**) is not characterized by high selectivity and sensitivity but can be easily implemented for many analytes, for example, Fe^{3+} ions. For obvious reasons, only a "turn off" response can be achieved for materials based on this principle. The IFE does not require chemical interaction of the analyte with the sensor material. The unequivocal evidence of the involvement of the IFE mechanism in the sensory response is the intersection of the spectra.

The second, also relatively simple, response mechanism involves either the coordination of the antenna molecule to the lanthanide complex or the elimination of the antenna molecule and its replacement with a non-antenna analyte molecule. In the first case, a turn on response is observed, and in the second, a turn off. Proof for this

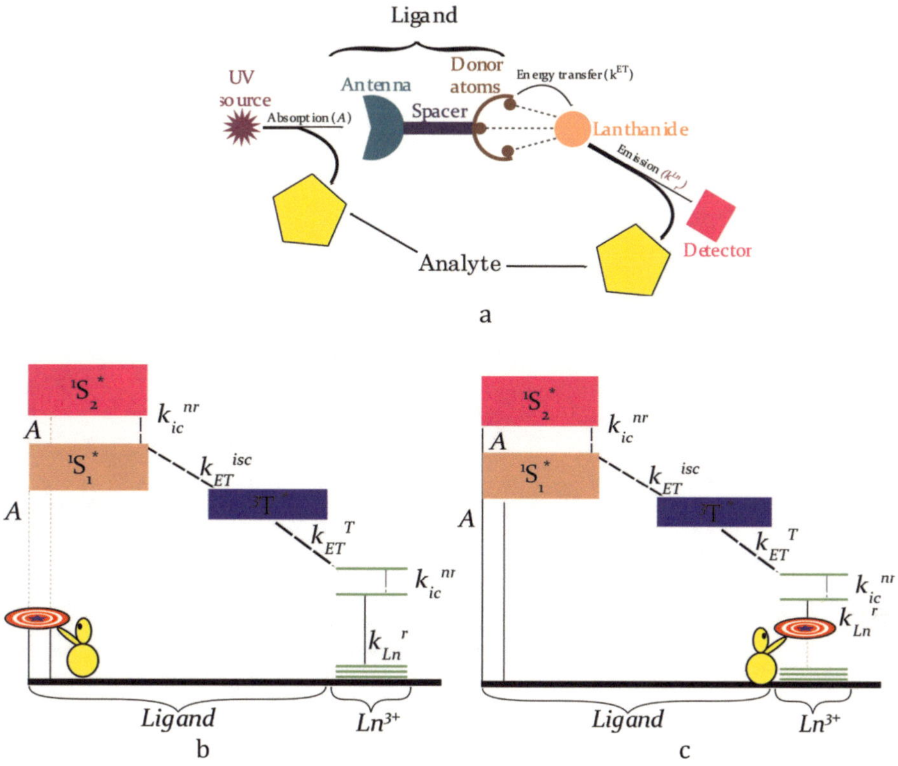

Figure 5.
Response mechanism associated with the internal filter effect: a) - the effect of the analyte on the luminescence of the sensor molecule; b) Jabłoński diagram when the analyte absorbs excitation energy; c) Jabłoński diagram for the absorption of the radiation emitted by the analyte.

mechanism can derive from analytical and spectroscopical data confirming the change in the structure of the complex.

The third possible mechanism, **analyte switching antenna** (**ANA**) is associated with a change in the structure of the ligand, "turning on" or "turning off" the antenna function (**Figure 6a**). In this case, the analyte changes either the absorption efficiency of the ligand (A) (**Figure 6b**), or the energy of the singlet and triplet levels (S^*, T^*) (**Figure 6c**), or the sensitization efficiency ($k^T{}_{ET}$) (**Figure 6d**), but the analyte itself should not have absorption in the region of excitation or emission of the sensor and should not have a direct effect (antenna sensitization or quenching) on the Ln^{3+} ion, otherwise other possible mechanisms (effect of an internal filter, energy transfer, vibrational quenching of luminescence) take place in parallel with this mechanism or dominate it. In a typical case related to this mechanism, the sensor molecule contains an antenna group, a macrocyclic analyte receptor as a spacer, and a lanthanide ion in a suitable environment. The coordination of M^{n+} by a macrocyclic receptor leads to a change in the electronic structure of the ligand, although the cation itself (usually an alkali or alkaline earth metal) does not affect the luminescence of other Ln^{3+} compounds.

Depending on the nature of the substance, both "turn off" and "turn on" responses can be realized. It is not easy to prove the existence of an ANA mechanism; it is necessary to show the formation of a ligand-analyte bond by analytical methods and to detect a change in the efficiency of specific processes using spectroscopic, kinetic, and/or calculation methods.

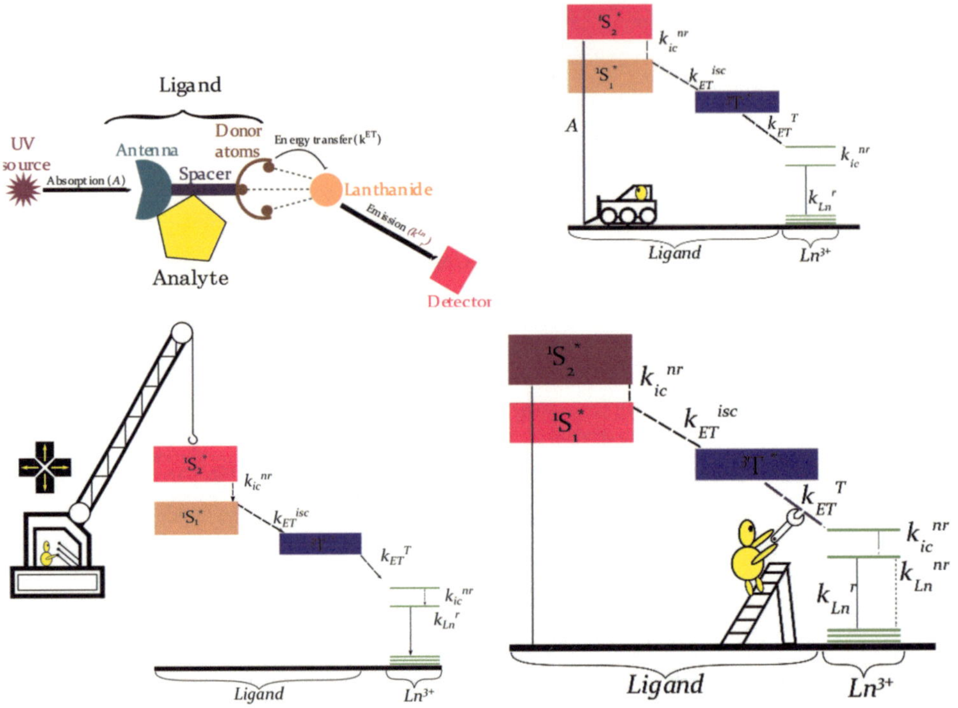

Figure 6.
Response mechanisms associated with an analyte-switched antenna (ANA). An analyte with its o antenna function is inserted into the structure of the complex (a). As a result, we have a change in the absorption efficiency (b), of the energy of the singlet or triplet level (c), and finally of the sensitization efficiency (d).

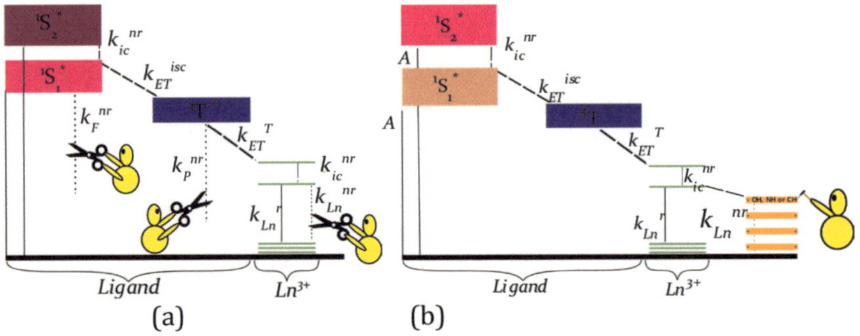

Figure 7.
Response mechanisms associated with the control of luminescence quenching processes (quencher addition/elimination, QAE). Suppression of nonradiative relaxation of the singlet, triplet states, and the excited state of lanthanide (a). Controlling vibrational quenching by introducing or removing X-H bonds (b).

The fourth possible mechanism is the addition or elimination (due to oxidation, substitution, or another process) of the quencher group during the interaction of the sensor with the analyte (quencher addition/elimination, QAE mechanism) (**Figure 7**). Bonds with suitable vibrational energy (OH, NH, CH) can act as quenching groups: in fact, the excited state of Ln^{3+} or the ligand upon interaction with the analyte can effectively undergo phonon relaxation. This mechanism involves the transformation of the sensor molecule into a new compound, which can be confirmed by analytical chemistry and spectroscopy. A "turn off" or a "turn on" response is observed depending on whether the groups are attached or split off. When a quenching group (for example, H_2O) is replaced by an antenna ligand, two possible processes simultaneously contribute: removal of the quencher and enhancement of sensitization by the new antenna molecule or the reverse process.

The fifth possible mechanism can be observed when the sensor material is a bimetallic system containing two lanthanum ions, usually Tb^{3+} and Eu^{3+}. Due to the proximity of radii and chemical properties, in these compounds, the two cations, as a rule, occupy the same crystallographic positions and are statistically distributed. If both Ln^{3+} ions are capable of emission, the excitation energy is transferred between the ions, and the rate constant of this process can be determined from kinetic data [23]. If the analyte affects the energy transfer constant between the ions, then a sensory response appears, which manifests itself as a change in the ratio of the integral luminescence intensities of the lanthanum ions. The proof of this mechanism requires a careful analysis of spectroscopic and kinetic data and cannot be extended to any "ratiometric" sensors, since in many cases, at least in the case of detection of CH, OH, and NH bonds, the response is not associated to a change in the efficiency of sensitization of one lanthanide ion by another, but with more efficient quenching of the luminescence of one of the ions.

Finally, also materials that are potentially usable as thermometric sensors can be based on REE compounds. However, the discussion on this topic is beyond the scope of this review. Lanthanide-containing luminescent thermometers are the subject of special reviews and monographs [24, 25].

The internal nature of the f-shells of lanthanides makes them less susceptible to the effects of the crystal field, however, a change in the geometry of the coordination environment can still be detected in several cases by careful analysis of the luminescence spectra. This principle is applicable in gauge sensors, in which compression of the crystal under pressure leads to a reduction in the Ln-O [26, 27] or Ln-F [28, 29] distances, which shifts the bands in the emission spectrum.

4. Turn off sensors

Recent papers describing luminescent sensor materials have reported clearly that "turn off sensors" dominate other types of response. This can be easily explained by the fact that luminescence quenching due to IFE, nonradiative relaxation of the excited states of the complex, or formation of a not-luminescent complex (static quenching) are caused by a large number of analytes with a different chemical nature [30]. Turn off sensors are poorly usable in real devices since it is difficult to detect signals at a low analyte concentration.

Effective internal filters can be either intensely colored compounds (due to d-d transitions typical of d-cations as Cu^{2+} or Co^{2+}, or to charge-transfer bands as in hydrolyzed Fe^{3+} aqua ions, $Cr_2O_7^{2-}$, CrO_4^{2-}, MnO_4^- anions).

The nonradiative relaxation of the excited state of the complex obeys the Stern-Volmer equation:

$$\frac{I_0}{I} = \frac{PLQY_0}{PLQY} = \frac{\tau_0}{\tau} = 1 + K_{SV}^D[Q]$$

where $K^D{}_{SV}$ – is the dynamic Stern-Volmer constant [31]. The higher the $K^D{}_{SV}$ value, the higher the sensitivity of the sensor material with respect to the quenching analyte. In the case of dynamic quenching, in addition to the luminescence intensity (and quantum yield), the lifetime of the excited state of Ln^{3+} is significantly reduced. Dynamic quenching is caused by the excitation energy adsorption from the quencher molecule without the formation of a phosphor-quencher bond.

Static quenching is caused by the formation of a non-luminescent complex formed by the luminophore and quencher molecules. The bond between the two species can be either covalent or hydrogen or stacking interaction. Static quenching is associated with nonradiative deactivation of the singlet or triplet state of the ligand, which weakens the antenna sensitization. As a result, during static quenching, the observed lifetime of the excited state (τ_{obs}) of the lanthanide ion can slightly decrease, but to a lesser extent with respect to the dynamic quenching. For static quenching, the Stern-Volmer equation is also applicable:

$$\frac{I_0}{I} = 1 + K^S_{SV}[Q]$$

In many cases, these two types of quenching complement each other, and by combining the equations for the static and dynamic cases the following generalized equation can be derived:

$$\frac{I_0}{I} = 1 + \left(K^D_{SV} + K^S_{SV}\right)Q + K^D_{SV} \times K^S_{SV} \times Q^2$$

The nonlinear dependence of I_0/I on the quencher concentration Q indicates a complex quenching mechanism.

Measurements carried out with a temperature control also make it possible to distinguish between static and dynamic quenching mechanisms, since for a static mechanism, an increase in temperature leads to a decrease in K_{SV} and for a dynamic one, to an increase [31].

The luminescence quenching can occur via excitation energy transfer (ET) from the ligand or metal through FRET (Förster Resonance Energy Transfer), DEE (Dexter Electron Exchange), or PET (Photoinduced Electron Transfer) mechanisms [32] (**Figure 8**) which imply the deactivation of the excited state of the molecule by transferring the excitation energy to the quencher molecule, whose LUMO is lower in energy with respect to the excited electron of the donor molecule.

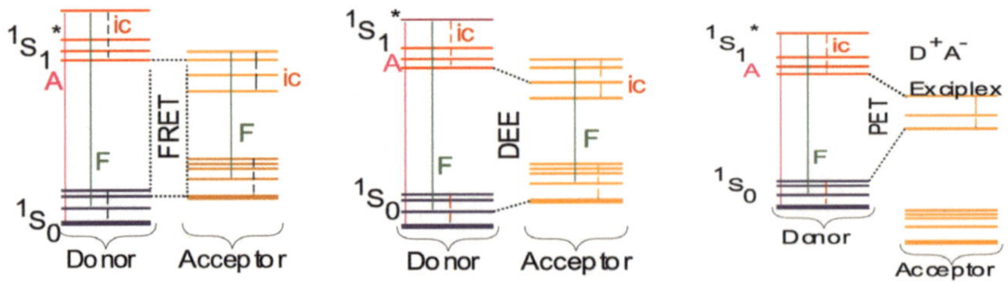

Figure 8.
Schematic representation of the mechanisms of deactivation of excited states. Left - Förster mechanism (FRET), center – Dexter mechanism (DEE), right – photoinduced electron transfer (PET) mechanism.

DOI: http://dx.doi.org/10.5772/intechopen.109011

FRET [33] implies nonradiative energy transfer from an excited donor molecule to an unexcited acceptor molecule, followed by radiative or nonradiative relaxation of the latter. The typical distances between the donor and acceptor molecules usually are 20–60 Å, and the efficiency of the Förster transfer decreases very rapidly with increasing distance (efficiency is proportional to r^{-6}), which makes it possible to measure the distance between particles. A transfer is described as the emission and absorption of a virtual photon [34].

DEE [35] includes two electronic transitions: an excited electron from the donor molecule to the LUMO of the acceptor molecule, and an electron from the HOMO of the donor to the HOMO of the acceptor. This creates a hole in the HOMO of the donor molecule [36]. These two processes proceed either simultaneously or sequentially. This mechanism is observed at much shorter distances (less than 10 Å) and is associated with the need for direct overlapping of the wave functions of the fluorophore and quencher. The efficiency of DEE quenching also decreases very rapidly with increasing distance (proportional to e^{-r}).

Another common mechanism of nonradiative relaxation of an excited state is PET [37] involving the formation of a charge-transfer complex (exciplex). An exciplex is formed when an electron moves from a LUMO donor to a LUMO acceptor, followed by radiative or nonradiative relaxation of the latter. An increase in the distance between the donor and acceptor has a much smaller effect on the efficiency of PET quenching with respect to FRET and DEE mechanisms.

To distinguish between these mechanisms, special quantum chemical methods are used to determine the HOMO and LUMO energies, as well as the dependence of the quenching efficiency on the distance between the donor and acceptor. Often, the real quenching mechanism has not been studied in detail, the PET or FRET quenching being only postulated.

The most common analytes detected with lanthanide-containing "turn off" sensors are listed in **Table 3**. Data on the mechanisms of sensory response are not always reported, moreover, in some cases, the author's interpretation of the mechanism may not be entirely correct. In several cases a static quenching mechanism associated with analyte-sensor bonds due to hydrogen bonds, stacking, and other weak interactions is suggested. Unfortunately, in most works, a more detailed mechanism of such static quenching has not been reported.

Analyte	Mechanism	Ref.
i. id-metal cations		
Cr^{2+}	Static quenching	[38]
Cr^{3+}	ET, IFE	[39, 40]
Mn^{2+}	Shift of the triplet level of the ligand during the coordination of the analyte	[41]
Fe^{2+}	Static quenching	[42]
Fe^{3+}	IFE, ET, Static quenching	[43–87]
Co^{2+}	ET, IFE, Static quenching	[88, 89]
Ni^{2+}	IFE, LMET	[90]
Cu^{2+}	IFE, LMET	[63, 69, 88, 91–99]

Analyte	Mechanism	Ref.
Ag^+	Shift of the triplet level of the ligand during the coordination of the analyte	[41, 66]
Hg^{2+}	IFE [100], Shift of the triplet level of the ligand during the coordination of the analyte, Static quenching [101, 102]	[100–102]
ii. p-metal cations		
Pb^{2+}	Shift of the triplet level of the ligand during the coordination of the analyte Static quenching [102, 103]	[102–104]
Al^{3+}	n/a,	[105, 106]
Ce^{3+}	IFE, Static quenching	[74]
UO_2^{2+}	n/a	[87]
iii. d-metal anions		
MnO_4^-	IFE, FRET	[106–108]
CrO_4^{2-}, $Cr_2O_7^{2-}$	IFE, FRET	[55, 77, 94, 105, 107–112]
ClO^-	Destruction of the ligand, weakening the antenna function	[113]
F^-	Displacement of the ligand by the analyte	[114]
HSO_4^-	Static quenching due to the H-bonding	[115]
$H_2PO_4^-$	Static quenching due to the H-bonding	[115]
CN^-	Static quenching due to the H-bonding. Possible destruction of the complex due to the binding of Zn^{2+} ions by CN^-	[116]
OAc^-	Static quenching due to the H-bonding	[116]
iv. Other organic analytes		
Nitroaromatic compounds	PET and IFE	[44, 46, 47, 67–71, 73, 75, 76, 85, 90, 94, 97, 99, 117–130]
Nitromethane	IFE	[53, 73]
$PhNH_2$	Static quenching due to the H-bonding и IFE	[131]
PhCHO	Static quenching due to the H-bonding	[40, 50, 132–134]
Chlorobenzenes	IFE	[135]
Acetone	Static quenching due to the H-bonding	[52, 54, 91, 99]
DMAA	n/a	[136]
Tryptophan	n/a	[59]
Tiaminphosphate derivatives	ET	[137, 138]
Acetophenone	n/a, but not the IFE	[139]
1-hydroxypyrene	Static quenching due to the H-bonding	[140]
Pesticides	Static quenching, IFE, FRET	[141]
Sulfamethazine	IFE	[142]
Ornidazole, Ronidazole	IFE&Energy transfer	[66]

Analyte	Mechanism	Ref.
Adenosine diphosphate	Static quenching	[115]
Quercetin	IFE	[62]
Nitrofuranose, Nitrofurantoin	PET IFE	[143]
Phenylglyoxylic acid	IFE	[144]

Table 3.
Proposed sensory response mechanism in turn off sensors employed to detect different analytes.

It should be noted that the most common mechanism is IFE. Typical excitation spectra of luminescent lanthanide complexes lie in the near UV region in the range 250–400 nm [4, 6, 145]. In this region, intense absorption of several types of analytes takes place, which makes such materials nonselective, although the selectivity has not been investigated in most of the articles.

The most efficient materials for the detection of various analytes are listed in **Table 4**. Materials with high K_{sv} values (>30,000) are selected as illustrative examples, making these materials potentially promising for practical applications.

Material	Media	Linearity range	K_{SV}. M^{-1}	LOD	Response time	Ref.
Fe^{3+} sensors						
Tb-MOF	DMF	0–100 μM	43,000	0.5 μM		[73]
Eu-MOF	DMF	0–140 μM	51,600	0.5 μM		[73]
Tb-MOF	EtOH	0.01–100 μM	114,300			[64]
Eu-MOF	EtOH	0–10 mM	43,000	1 μM	8 min	[63]
Tb-MOF	H_2O		65,000	0.84 μM		[57]
Tb-MOF	H_2O	0–1 mM	96,021		24 h	
Eu-MOF	H_2O	0–1 mM	75,596		24 h	[52]
Eu-MOF	H_2O	5–12 μM	119,000	0.277 μM		[66]
Cu^{2+} sensors						
Nd-MOF	CH_3CN	0–100 μM	80,000	0.39 μM		[88]
Eu-MOF	EtOH	0–10 mM	52,000	1 μM	8 min	[63]
Cr^{2+} sensors						
Eu-MOF	H_2O	0–100 μM	39,000			[38]
Tb-MOF	H_2O	0–100 μM	50,000			[38]
Co^{2+} sensor						
Nd-MOF	CH_3CN	0–60 μM	66,000	0.47 μM		[88]
Ag^{+} sensor						
Eu-MOF	H_2O	10–19 μM	118,000	0.272 μM		[66]

Material	Media	Linearity range	K_{SV}. M^{-1}	LOD	Response time	Ref.
Pb^{2+} sensor						
Eu-MOF	H_2O DMF	10–500 μM	33,014 43,988	91 μM 68 μM		[102]
$Cr_2O_7^{2-}$ sensors						
POSS-PVIM@[Eu(dbm)$_3$] hybrid material	H_2O	5–40 μM	42,014	0.68 μM	5 min	[111]
[Eu-MOF	H_2O	0–200 μM	220,000		6 min	[55]
Tb-MOF	H_2O	0–100 μM	41,100	5.6 ppb	5 min	[112]
Tb-MOF	H_2O	0–3 μM	106,500	0.32 μM		[77]
Eu-MOF	H_2O	0–1.5 μM	151,000	0.10 μM		[77]
CrO_4^{2-} sensors						
Tb-MOF	H_2O	0–3 μM	127,500	0.33 μM		[77]
Eu-MOF	H_2O	0–2 μM	102,500	0.17 μM		[77]
TNP sensors						
La-MOF	H_2O	0–20 nM	2.4*109	0.22 ppb		[129]
Nd-MOF	H_2O	0–20 nM	1*107	0.71 ppb		
Pr-MOF	H_2O	0–20 nM	6.8*109	0.27 ppb		
Eu-MOF	DMAA	0–25 μM	49,000	0.34 μM		[130]
Tb-MOF	DMF	0–22 μM	155,000	30 nM	30 min	[71]
Tb-MOF	H_2O	0–40 μM	54,800	0.35 ppm		[118]
Nitrobenzene sensors						
Tb-MOF	H_2O	5–30 ppm	93,400	2.1 ppm		[99]
	H_2O	5–30 ppm	80,300	2.2 ppm		[99]
Eu-MOF	DMF	0–1.6 mM	62,400	5 μM		[119]
$PhNH_2$ sensor						
Eu(NO_3)$_3$Zn-MOF	DMF	0–1 μM	9,390,000	6.6 nM	30 min	[110]
PhCHO sensors						
Tb(NO_3)$_3$ + Zn-MOF	DMF	0–0.2 μM	6,170,000	10 nM	30 min	[110]
Tb-MOF	EtOH	0–0.5%	31,100			[133]
Tb-MOF	EtOH	0–20 μM	33,500	$7.71e^{-7}$ M	30 min	[134]
Yb-MOF	EtOH	0–20 μM	29,900	$8.64e^{-7}$ M	30 min	[134]
Acetone sensor						
Eu-MOF	DMF	5–30 ppm	78,900	10.1 ppm		[99]
Tiamine phosphates chlorides						
Eu-MOF			Up to 900,800	Up to 0.029 μM		[137]
Tb-MOF			Up to 39,000	Up to 0.025 μM		[138]

Material	Media	Linearity range	K_{SV}, M^{-1}	LOD	Response time	Ref.
Sulfamethazine sensor						
Eu-MOF	EtOH-H_2O	0–100 μM	45,980	0.66 μM		[142]
Ornidazole sensor						
Eu-MOF	H_2O	6–18 μM	197,000	0.183 μM		[66]
Ronidazole sensor						
Eu-MOF	H_2O	6–18 μM	212,000	0. 141 μM		[66]

Table 4.
The most sensitive "turn off" touch materials.

The detection of Fe^{3+}, d-metal ions and also d-metal-containing anions with strong oxidizing abilities can be performed by simpler methods with respect to the production of the luminescent sensors. At the same time, the determination of impurities of toxic metals (lead, mercury) in waters or foods seems to be a much more urgent task, and there is hope that other results on the determination of such analytes will appear soon. Nitroaromatic compounds are found in a large number of explosives and ammunition, and proposed sensor materials often exhibit very high sensitivity to them. At present, there are some commercial luminescent "turn off" detectors of nitroaromatic compounds, and although trade secrets prevent disclosure of the compositions of specific materials, lanthanide-containing complexes seem to be promising in this direction. As for the detection of other organic molecules, due to the prevalence of absorption in the near UV region, the industrial production of these materials requires increased selectivity.

Based on the analysis of literature data, we can propose the following algorithm for establishing the sensory response mechanism for "turn off" sensors:

i. A plot of Stern-Volmer graphs at different temperatures to establish the presence of dynamic or static quenching or their simultaneous presence. An exclusively static quenching mechanism indicates the formation of a not-luminescent complex, which, however, requires additional evidence.

ii. A chemical and/or phase analysis of the "sensor-analyte" system, is absolutely necessary to verify the formation of a not-luminescent complex and the destruction of the sensor structure.

iii. A sensory experiment in a deuterated solvent medium to determine if the material is sensitive to water, alcohol, and other molecules with an active OH group. The disappearance of the response unambiguously indicates that OH quenching is the dominant mechanism.

iv. A study of the absorption spectra of the analyte and the excitation spectra of the sensor material. An overlap suggests at least partial involvement of the IFE mechanism. Measurements of the sensory response should be made when the material is excited at a wavelength that does not fall within the absorption spectrum of the sensor, for example, directly at the f-f wavelength of the Ln^{3+}

transition. If the response completely disappears, then only the IFE mechanism is in place.

v. A study of the luminescence spectra of the complex and absorption spectra of the analyte. The presence of overlap indicates the partial participation of the IFE mechanism and the possibility of energy transfer through the FRET mechanism.

vi. A quantum-chemical modeling study to determine the energy of orbitals, measurement of τ upon excitation through the bands of the ligand and of metal in the absence and presence of quenchers. In this way, the presence of energy transfer can be confirmed and the specific mechanism (PET, FRET, DEE) can be determined.

5. Conclusions

The luminescence of lanthanide coordination compounds is an unusual and multi-stage process that imposes numerous restrictions on the structure of both ligands and complex compounds. Interruption of any of the intermediate stages due to the *effect of an internal filter*, *nonradiative deactivation* of the excited states of the ligand and metal, *changes in the energy of the triplet level* of the ligand, or other factors dramatically affect the luminescence brightness. This presents a serious challenge in the development of efficient luminescent materials for light-emitting devices, but it can be an elegant key for the development of sensitive sensor materials for the detection of analytes of various natures.

The main problem with *turn off* materials is *low selectivity*, since dissimilar analytes such as nitroaromatic compounds and Fe^{3+} cations, for example, cause exactly the same response. Now, a new challenge for researchers is to find the way to increase the selectivity of the response, which requires the use of modern approaches of supramolecular chemistry and crystal chemical engineering.

In the following chapter, we will focus on consideration of two less common, but more technologically advanced and promising areas in lanthanide-based luminescent chemical sensors – *turn on* and *ratiometric* materials.

Acknowledgements

University of Camerino is gratefully acknowledged.

Conflict of interest

The authors declare no conflict of interest.

Author details

Claudio Pettinari[1]*, Andrei Drozdov[2] and Yuriy Belousov[2]

1 School of Pharmacy, University of Camerino, Camerino, Italy

2 Moscow State University, Moscow, Russia

*Address all correspondence to: claudio.pettinari@unicam.it

References

[1] Feng J, Zhang H. Hybrid materials based on lanthanide organic complexes: A review. Chemical Society Reviews. 2013;**42**:387-410. DOI: 10.1039/c2cs35069f

[2] Sun T, Gao Y, Du Y, Zhou L, Chen X. Recent advances in developing lanthanide metal–organic frameworks for ratiometric fluorescent sensing. Frontiers in Chemistry. 2021;**8**. DOI: 10.3389/fchem.2020.624592. Available from: https://www.frontiersin.org/articles/10.3389/fchem.2020.624592

[3] Bünzli JCG, Piguet C. Lanthanide-containing molecular and supramolecular polymetallic functional assemblies. Chemical Reviews. 2002;**102**: 1897-1928. DOI: 10.1021/cr010299j

[4] Bünzli JCG, Piguet C. Taking advantage of luminescent lanthanide ions. Chemical Society Reviews. 2005; **34**:1048-1077. DOI: 10.1039/b406082m

[5] Eliseeva SV, Bünzli JCG. Lanthanide luminescence for functional materials and bio-sciences. Chemical Society Reviews. 2010;**39**:189-227. DOI: 10.1039/b905604c

[6] Eliseeva SV, Bünzli JCG. Rare earths: Jewels for functional materials of the future. New Journal of Chemistry. 2011; **35**:1165-1176. DOI: 10.1039/c0nj00969e

[7] Lee SY, Lin M, Lee A, Park YII. Lanthanide-Doped Nanoparticles for Diagnostic Sensing. Nanomaterials. 2017;**7**:411. DOI: 10.3390/nano7120411

[8] Aulsebrook ML, Graham B, Grace MR, Tuck KL. Lanthanide complexes for luminescence-based sensing of low molecular weight analytes. Coordination Chemistry Reviews. 2018;**375**:191-220

[9] Aletti AB, Gillen DM, Gunnlaugsson T. Luminescent/colorimetric probes and (chemo-) sensors for detecting anions based on transition and lanthanide ion receptor/binding complexes. Coordination Chemistry Reviews. 2018;**354**:98-120

[10] Xu H, Cao CS, Kang XM, Zhao B. Lanthanide-based metal-organic frameworks as luminescent probes. Dalton Transactions. 2016;**45**:18003-18017. DOI: 10.1039/C6DT02213H

[11] Zhao SN, Wang G, Poelman D, Voort PV. Luminescent Lanthanide MOFs: A Unique Platform for Chemical Sensing. Materials. Basel, Switzerland. 2018;**11**(4): 572. DOI: 10.3390/ma11040572

[12] Lunev AM, Belousov YA. Luminescent sensor materials based on rare-earth element complexes for detecting cations, anions, and small molecules. Russian Chemical Bulletin. 2022;**71**:825-857. DOI: 10.1007/s11172-022-3485-3

[13] Van Der Ende BM, Aarts L, Meijerink A. Lanthanide ions as spectral converters for solar cells. Physical Chemistry Chemical Physics. 2009;**11**: 11081-11095. DOI: 10.1039/b913877c

[14] Lis S, Kimura T, Yoshida Z. Luminescence lifetime of lanthanide(III) ions in aqueous solution containing azide ion. Journal of Alloys and Compounds. 2001;**323–324**:125-127. DOI: 10.1016/S0925-8388(01)00980-X

[15] Poluektov NS, Kononeko LI, Efrushina NP, Beltukova SV. In: Pilipenko AT, editor. Spectrophotometric and luminescent methods for the determination of lanthanides. Kiev, USSR: Naukova Dumka; 1989;**256**

[16] Weissman SI. Intramolecular energy transfer the fluorescence of complexes of europium. The Journal of Chemical Physics. 1942;**10**:214-217. DOI: 10.1063/1.1723709

[17] Sato S, Wada M. Relations between intramolecular energy transfer efficiencies and triplet state energies in rare earth β-Diketone chelates. Bulletin of the Chemical Society of Japan. 1970; **43**:1955-1962

[18] Bünzli, J.-C.G.; Eliseeva, S. V. Basics of lanthanide photophysics. 2010; pp. 1–45

[19] Latva M, Takalob H, Mukkala VM, Matachescu C, Rodríguez-Ubis JC, Kankare J. Correlation between the lowest triplet state energy level of the ligand and lanthanide(III) luminescence quantum yield. Journal of Luminescence. 1997;**75**:149-169. DOI: 10.1016/S0022-2313(97)00113-0

[20] Bünzli JCG. Lanthanide luminescence: From a mystery to rationalization, understanding, and applications. Handbook on the Physics and Chemistry of Rare Earths. 2016;**50**: 141-176. DOI: 10.1016/bs.hpcre.2016.08.003

[21] Bünzli J, Eliseeva SV. Basics of lanthanides photophysics. In: Hänninen P, Lanthan HH, editors. Lanthanide Luminescence. Vol. 7. Lumin. Springer Ser. Fluoresc; 2010. pp. 1-45. DOI: 10.1007/4243

[22] Hu Z, Deibert BJ, Li J. Luminescent metal-organic frameworks for chemical sensing and explosive detection. Chemical Society Reviews. 2014;**43**: 5815-5840

[23] Gontcharenko VE, Kiskin MA, Dolzhenko VD, Korshunov VM, Taydakov IV, Belousov YA. Mono- and mixed metal complexes of Eu^{3+}, Gd^{3+}, and Tb^{3+} with a diketone, bearing pyrazole moiety and chf_2-group: Structure, color tuning, and kinetics of energy transfer between lanthanide ions. Molecules. 2021;**26**:1-16. DOI: 10.3390/molecules26092655

[24] Brites CDS, Millán A, Carlos LD. Lanthanides in luminescent thermometry. In: Bünzli J-C, Pecharsky V, editors. Handbook on the Physics and Chemistry of Rare Earths. Vol. 49. Elsevier B.V; 2016. pp. 339-427. DOI: 10.1016/bs.hpcre.2016.03.005. ISSN: 0168-1273; ISBN: 9780444636997

[25] Rocha J, Brites CDS, Carlos LD. Lanthanide organic framework luminescent thermometers. Chemistry-A European Journal. 2016;**22**: 14782-14795. DOI: 10.1002/chem.201600860

[26] Runowski M, Woźny P, Martín IR. Optical pressure sensing in vacuum and high-pressure ranges using lanthanide-based luminescent thermometer-manometer. Journal of Materials Chemistry C. 2021;**9**:4643-4651. DOI: 10.1039/d1tc00709b

[27] Runowski M, Woźny P, Lavín V, Lis S. Optical pressure nano-sensor based on lanthanide doped SrB2O4:Sm2+ luminescence – novel high-pressure nanomanometer. Sensors and Actuators B: Chemical. 2018;**273**:585-591. DOI: 10.1016/j.snb.2018.06.089

[28] Goderski S, Runowski M, Woźny P, Lavín V, Lis S. Lanthanide upconverted luminescence for simultaneous contactless optical thermometry and manometry-sensing under extreme conditions of pressure and temperature. ACS Applied Materials & Interfaces. 2020;**12**:40475-40485. DOI: 10.1021/acsami.0c09882

[29] Antoniak MA, Zelewski SJ, Oliva R, Żak A, Kudrawiec R, Nyk M. Combined temperature and pressure sensing using luminescent NaBiF4:Yb,Er nanoparticles. ACS Applied Nano Materials. 2020;**3**:4209-4217. DOI: 10.1021/acsanm.0c00403

[30] Yan B. Luminescence response mode and chemical sensing mechanism for lanthanide-functionalized metal-organic framework hybrids. Inorganic Chemistry Frontiers. 2021;**8**:201-233

[31] Lakowicz JR. In: Lakowicz JR, editor. Principles of Fluorescence Spectroscopy. Boston, MA: Springer US; 2006

[32] Dutta A, Singh A, Wang X, Kumar A, Liu J. Luminescent sensing of nitroaromatics by crystalline porous materials. CrystEngComm. 2020;**22**: 7736-7781. DOI: 10.1039/d0ce01087a

[33] Förster T. Zwischenmolekulare energiewanderung und fluoreszenz. Annals of Physics. 1948;**437**:55-75. DOI: 10.1002/andp.19484370105

[34] Jones GA, Bradshaw DS. Resonance energy transfer: From fundamental theory to recent applications. Frontiers of Physics. 2019;**7**:1-19. DOI: 10.3389/fphy.2019.00100

[35] Dexter DL. A theory of sensitized luminescence in solids. The Journal of Chemical Physics. 1953;**21**:836-850. DOI: 10.1063/1.1699044

[36] Yamase, T.; 2001. pp. 187–203

[37] Fox MA. Photoinduced electron transfer. Photochemistry and Photobiology. 1990;**52**:617-627. DOI: 10.1111/j.1751-1097.1990.tb01808.x

[38] Huang XH, Shi L, Ying SM, Yan GY, Liu LH, Sun YQ, et al. Two lanthanide metal-organic frameworks as sensitive luminescent sensors for the detection of Cr2+ and Cr2O72- in aqueous solutions. CrystEngComm. 2018;**20**:189-197. DOI: 10.1039/c7ce01781b

[39] Zheng K, Liu Z, Jiang Y, Guo P, Li H, Zeng C, et al. Ultrahigh luminescence quantum yield lanthanide coordination polymer as a multifunctional sensor. Dalton Transactions. 2018;**47**: 17432-17440. DOI: 10.1039/c8dt03832e

[40] Sun Z, Yang M, Ma Y, Li L. Multi-responsive luminescent sensors based on two-dimensional lanthanide-metal organic frameworks for highly selective and sensitive detection of Cr (III) and Cr(VI) ions and benzaldehyde. Crystal Growth & Design. 2017;**17**: 4326-4335. DOI: 10.1021/acs.cgd.7b00638

[41] Li YX, Li SJ, Yan PF, Wang XY, Yao X, An GH, et al. Luminescence-colour-changing sensing of Mn2+ and Ag+ ions based on a white-light-emitting lanthanide coordination polymer. Chemical Communications. 2017;**53**:5067-5070. DOI: 10.1039/c7cc00258k

[42] Meng K, Yao C, Ma Q, Xue Z, Du Y, Liu W, et al. A reversibly responsive fluorochromic hydrogel based on lanthanide–mannose complex. Advancement of Science. 2019;**6**. DOI: 10.1002/advs.201802112. Article ID: 1802112

[43] Gu JZ, Cai Y, Liu Y, Liang XX, Kirillov AM. New lanthanide 2D coordination polymers constructed from a flexible ether-bridged tricarboxylate block: Synthesis, structures and luminescence sensing. Inorganica Chimica Acta. 2018;**469**:98-104. DOI: 10.1016/j.ica.2017.08.054

[44] Sun Z, Li Y, Ma Y, Li L. Dual-functional recyclable luminescent sensors based on 2D lanthanide-based metal-organic frameworks for highly sensitive detection of Fe^{3+} and 2,4-dinitrophenol. Dyes and Pigments. 2017; **146**:263-271. DOI: 10.1016/j.dyepig.2017.07.015

[45] Liu L, Wang Y, Lin R, Yao Z, Lin Q, Wang L, et al. Two water-stable lanthanide metal-organic frameworks with oxygen-rich channels for fluorescence sensing of Fe(Iii) ions in aqueous solution. Dalton Transactions. 2018;**47**:16190-16196. DOI: 10.1039/C8DT03741H

[46] Ma JJ, Liu WS. Effective luminescence sensing of Fe3+, Cr2O72-, MnO4- and 4-nitrophenol by lanthanide metal-organic frameworks with a new topology type. Dalton Transactions. 2019;**48**:12287-12295. DOI: 10.1039/c9dt01907c

[47] Feng X, Shang Y, Zhang H, Liu X, Wang X, Chen N, et al. Multi-functional lanthanide-CPs based on tricarboxylphenyl terpyridyl ligand as ratiometric luminescent thermometer and highly sensitive ion sensor with turn on/off effect. Dalton Transactions. 2020; **49**:4741-4750. DOI: 10.1039/d0dt00310g

[48] Jin J, Yang G, Liu Y, Cheng S, Liu J, Wu D, et al. Two series of microporous lanthanide-organic frameworks with different secondary building units and exposed lewis base active sites: Sensing, dye adsorption, and magnetic properties. Inorganic Chemistry. 2019;**58**:339-348. DOI: 10.1021/acs.inorgchem.8b02435

[49] Jing T, Chen L, Jiang F, Yang Y, Zhou K, Yu M, et al. Fabrication of a robust lanthanide metal-organic framework as a multifunctional material for Fe(III) detection, CO2 capture, and utilization. Crystal Growth & Design. 2018;**18**:2956-2963. DOI: 10.1021/acs.cgd.8b00068

[50] Li R, Qu XL, Zhang YH, Han HL, Li X. Lanthanide-organic frameworks constructed from naphthalenedisulfonates: Structure, luminescence and luminescence sensing properties. CrystEngComm. 2016;**18**: 5890-5900. DOI: 10.1039/c6ce01028h

[51] Li Y, Xu Y, Wang Y. Preparation and properties of transparent ultrathin lanthanide-complex films. Chemistry-A European Journal. 2016;**22**:10976-10982. DOI: 10.1002/chem.201601189

[52] Liu F, Gao W, Li P, Zhang XM, Liu JP. Lanthanide metal-organic frameworks as multifunctional luminescent sensor for detecting cations, anions and organic solvent molecules in aqueous solution. Journal of Solid State Chemistry. 2017;**253**:202-210. DOI: 10.1016/j.jssc.2017.05.040

[53] Ma L, Zhang Q, Wu H, Yang J, Liu YY, Ma JF. Multifunctional luminescence sensors assembled with lanthanide and a cyclotriveratrylene-based ligand. European Journal of Inorganic Chemistry. 2017;**2017**: 4221-4230. DOI: 10.1002/ejic.201700874

[54] Wang J, Yu M, Chen L, Li Z, Li S, Jiang F, et al. Construction of a stable lanthanide metal-organic framework as a luminescent probe for rapid naked-eye recognition of Fe^{3+} and acetone. Molecules. 2021;**26**:1695. DOI: 10.3390/molecules26061695

[55] Li GP, Liu G, Li YZ, Hou L, Wang YY, Zhu Z. Uncommon pyrazoyl-carboxyl bifunctional ligand-based microporous lanthanide systems: Sorption and luminescent sensing properties. Inorganic Chemistry. 2016;

55:3952-3959. DOI: 10.1021/acs.inorgchem.6b00217

[56] Wu YP, Xu GW, Dong WW, Zhao J, Li DS, Zhang J, et al. Anionic lanthanide MOFs as a platform for iron-selective sensing, systematic color tuning, and efficient nanoparticle catalysis. Inorganic Chemistry. 2017;**56**:1402-1411. DOI: 10.1021/acs.inorgchem.6b02476

[57] Yang D, Lu L, Feng S, Zhu M. First Ln-MOF as a trifunctional luminescent probe for the efficient sensing of aspartic acid, Fe3+and DMSO. Dalton Transactions. 2020;**49**:7514-7524. DOI: 10.1039/d0dt00938e

[58] Jia P, Wang Z, Zhang Y, Zhang D, Gao W, Su Y, et al. Selective sensing of Fe3+ ions in aqueous solution by a biodegradable platform based lanthanide metal organic framework. Spectrochimica Acta, Part A: Molecular and Biomolecular Spectroscopy. 2020:230. DOI: 10.1016/j.saa.2020.118084

[59] Abdelhamid HN, Bermejo-Gómez A, Martín-Matute B, Zou X. A water-stable lanthanide metal-organic framework for fluorimetric detection of ferric ions and tryptophan. Microchimica Acta. 2017;**184**:3363-3371. DOI: 10.1007/s00604-017-2306-0

[60] Fan B, Wei J, Ma X, Bu X, Xing N, Pan Y, et al. Synthesis of lanthanide-based room temperature ionic liquids with strong luminescence and selective sensing of Fe(III) over mixed metal ions. Industrial and Engineering Chemistry Research. 2016;**55**:2267-2271. DOI: 10.1021/acs.iecr.5b03947

[61] Li L, Chen Q, Niu Z, Zhou X, Yang T, Huang W. Lanthanide metal-organic frameworks assembled from a fluorene-based ligand: selective sensing of Pb2+ and Fe3+ Ions. Journal of Materials Chemistry C. 2016;**4**:1900-1905. DOI: 10.1039/c5tc04320d

[62] Xu QW, Dong G, Cui R, Li X. 3D lanthanide-coordination frameworks constructed by a ternary mixed-ligand: Crystal structure, Luminescence and luminescence sensing. CrystEngComm. 2020;**22**:740-750. DOI: 10.1039/c9ce01779h

[63] Zhang H, Fan R, Chen W, Fan J, Dong Y, Song Y, et al. 3D lanthanide metal-organic frameworks based on mono-, tri-, and heterometallic tetranuclear clusters as highly selective and sensitive luminescent sensor for Fe3+ and Cu2+ ions. Crystal Growth & Design. 2016;**16**:5429-5440. DOI: 10.1021/acs.cgd.6b00903

[64] Wang KM, Du L, Ma YL, Zhao JS, Wang Q, Yan T, et al. Multifunctional chemical sensors and luminescent thermometers based on lanthanide metal-organic framework materials. CrystEngComm. 2016;**18**:2690-2700. DOI: 10.1039/c5ce02367j

[65] Xu H, Hu HC, Cao CS, Zhao B. Lanthanide organic framework as a regenerable luminescent probe for Fe3+. Inorganic Chemistry. 2015;**54**(10): 4585-4587. DOI: 10.1021/acs.inorgchem.5b00113

[66] Li JM, Huo R, Li X, Sun HL. Lanthanide-organic frameworks constructed from 2,7-Naphthalenedisulfonate and 1 H-Imidazo[4,5- f][1,10]-phenanthroline: Synthesis, structure, and luminescence with near-visible light excitation and magnetic properties. Inorganic Chemistry. 2019;**58**:9855-9865. DOI: 10.1021/acs.inorgchem.9b00925

[67] Wang D, Sun L, Hao C, Yan Y, Liang Z. Lanthanide metal-organic

frameworks based on a 1,2,3-Triazole-containing tricarboxylic acid ligand for luminescence sensing of metal ions and nitroaromatic compounds. RSC Advances. 2016;**6**:57828-57834. DOI: 10.1039/c6ra06303a

[68] Li ZJ, Li XY, Yan YT, Hou L, Zhang WY, Wang YY. Tunable emission and selective luminescence sensing in a series of lanthanide metal-organic frameworks with uncoordinated lewis basic triazolyl sites. Crystal Growth & Design. 2018;**18**:2031-2039. DOI: 10.1021/acs.cgd.7b01453

[69] Smith JA, Singh-Wilmot MA, Carter KP, Cahill CL, Ridenour JA. Lanthanide-2,3,5,6-tetrabromoterephthalic acid metal-organic frameworks: Evolution of halogen·halogen interactions across the lanthanide series and their potential as selective bifunctional sensors for the detection of Fe 3+, Cu 2+, and nitroaromatics. Crystal Growth & Design. 2019;**19**: 305-319. DOI: 10.1021/acs.cgd.8b01426

[70] Cao XM, Wei N, Liu L, Li L, Han ZB. Luminescent lanthanide-organic polyrotaxane framework as a turn-off sensor for nitrobenzene and Fe3+. RSC Advances. 2016;**6**:19459-19462. DOI: 10.1039/c5ra25872c

[71] Zhang X, Zhan Z, Liang X, Chen C, Liu X, Jia Y, et al. Lanthanide-MOFs constructed from mixed dicarboxylate ligands as selective multi-responsive luminescent sensors. Dalton Transactions. 2018;**47**:3272-3282. DOI: 10.1039/c7dt02966g

[72] Liu W, Huang X, Xu C, Chen C, Yang L, Dou W, et al. A multi-responsive regenerable europium–organic framework luminescent sensor for Fe3+, CrVIAnions, and picric acid. Chemistry —A European Journal. 2016;**22**: 18769-18776. DOI: 10.1002/chem.201603607

[73] Yan W, Zhang C, Chen S, Han L, Zheng H. Two lanthanide metal–organic frameworks as remarkably selective and sensitive bifunctional luminescence sensor for metal ions and small organic molecules. ACS Applied Materials & Interfaces. 2017;**9**:1629-1634. DOI: 10.1021/acsami.6b14563

[74] Zhang Q, Wang J, Kirillov AM, Dou W, Xu C, Xu C, et al. Multifunctional Ln-MOF luminescent probe for efficient sensing of Fe3+, Ce3+, and acetone. ACS Applied Materials & Interfaces. 2018;**10**:23976-23986. DOI: 10.1021/acsami.8b06103

[75] Wen RM, Han S, de Ren GJ, Chang Z, Li YW, Bu XH. A flexible zwitterion ligand based lanthanide metal-organic framework for luminescence sensing of metal ions and small molecules. Dalton Transactions. 2015;**44**:10914-10917. DOI: 10.1039/c4dt02445a

[76] Lu SQ, Liu YY, Duan ZM, Wang ZX, Li MX, He X. Improving water-stability and porosity of lanthanide metal-organic frameworks by stepwise synthesis for sensing and removal of heavy metal ions. Crystal Growth & Design. 2018;**18**: 4602-4610. DOI: 10.1021/acs.cgd.8b00575

[77] Mi X, Sheng D, Yu Y, Wang Y, Zhao L, Lu J, et al. Tunable light emission and multiresponsive luminescent sensitivities in aqueous solutions of two series of lanthanide metal-organic frameworks based on structurally related ligands. ACS Applied Materials & Interfaces. 2019;**11**: 7914-7926. DOI: 10.1021/acsami.8b18320

[78] Silva Lins IM, da Silva D, Fonseca J, Lourenço da Luz L, Chojnacki J, Júnior SA, et al. Novel luminescent calixarene-based lanthanide materials: From synthesis and characterization to the selective detection of Fe3+. Journal of Solid State Chemistry. 2021:295. DOI: 10.1016/j.jssc.2020.121916

[79] Zheng K, Lou KL, Zeng CH, Li SS, Nie ZW, Zhong S. Hybrid membrane of agarose and lanthanide coordination polymer: A selective and sensitive Fe3+ sensor. Photochemistry and Photobiology. 2015;**91**:814-818. DOI: 10.1111/php.12460

[80] Wang Y, Huang R, Zhang J, Cheng G, Yang H. Lanthanide(Tb3+, Eu3+)-functionalized a new one dimensional Zn-MOF composite as luminescent probe for highly selectively sensing Fe3+. Polyhedron. 2018;**148**:178-183. DOI: 10.1016/j.poly.2018.04.013

[81] Fan W, Du J, Kou J, Zhang Z, Liu F. Hierarchical porous cellulose/lanthanide hybrid materials as luminescent sensor. Journal of Rare Earths. 2018;**36**: 1036-1043. DOI: 10.1016/j.jre.2018.03.021

[82] Liang YT, Yang GP, Liu B, Yan YT, Xi ZP, Wang YY. Four super water-stable lanthanide-organic frameworks with active uncoordinated carboxylic and pyridyl groups for selective luminescence sensing of Fe3+. Dalton Transactions. 2015;**44**:13325-13330. DOI: 10.1039/c5dt01421b

[83] Zhao XL, Tian D, Gao Q, Sun HW, Xu J, Bu XH. A chiral lanthanide metal-organic framework for selective sensing of Fe(III) Ions. Dalton Transactions. 2016;**45**:1040-1046. DOI: 10.1039/c5dt03283k

[84] Ning Y, Wang L, Yang GP, Wu Y, Bai N, Zhang W, et al. Four new lanthanide-organic frameworks: Selective luminescent sensing and magnetic properties. Dalton Transactions. 2016;**45**:12800-12806. DOI: 10.1039/c6dt01393g

[85] Chen M, Xu WM, Tian JY, Cui H, Zhang JX, Liu C, et al. A Terbium(Iii) lanthanide-organic framework as a platform for a recyclable multi-responsive luminescent sensor. Journal of Materials Chemistry C. 2017;**5**: 2015-2021. DOI: 10.1039/C6TC05615F

[86] Liu LH, Qiu XT, Wang YJ, Shi Q, Sun YQ, Chen YP. NIR emission and luminescent sensing of a lanthanide-organic framework with lewis basic imidazole and pyridyl sites. Dalton Transactions. 2017;**46**:12106-12113. DOI: 10.1039/c7dt02745a

[87] Wang Y, Xing SH, Bai FY, Xing YH, Sun LX. Stable lanthanide-organic framework materials constructed by a triazolyl carboxylate ligand: Multifunction detection and white luminescence tuning. Inorganic Chemistry. 2018;**57**:12850-12859. DOI: 10.1021/acs.inorgchem.8b02050

[88] Shi D, Yang X, Ma Y, Niu M, Jones RA. Construction of 14-metal lanthanide nanorings with NIR luminescence response to ions. Chemical Communications. 2020;**56**:8651-8654. DOI: 10.1039/d0cc04242k

[89] Cui Z, Zhang X, Liu S, Zhou L, Li W, Zhang J. Anionic lanthanide metal-organic frameworks: Selective separation of cationic dyes, solvatochromic behavior, and luminescent sensing of Co (II) ion. Inorganic Chemistry. 2018;**57**: 11463-11473. DOI: 10.1021/acs.inorgchem.8b01319

[90] Li JJ, Fan TT, Qu XL, Han HL, Li X. Temperature-induced 1D lanthanide polymeric frameworks based on Lnn

(n = 2, 2, 4, 6) cores: Synthesis, crystal structures and luminescence properties. Dalton Transactions. 2016;**45**:2924-2935. DOI: 10.1039/c5dt04262c

[91] Lian X, Yan B. Novel core-shell structure microspheres based on lanthanide complexes for white-light emission and fluorescence sensing. Dalton Transactions. 2016;**45**:2666-2673. DOI: 10.1039/c5dt03939h

[92] Su R, Gao J, Deng S, Zhang R, Zheng Y. Dual-target optical sensors assembled by lanthanide complex incorporated sol–gel-derived polymeric films. Journal of Sol-Gel Science and Technology. 2016;**78**:606-612. DOI: 10.1007/s10971-016-3982-7

[93] Liang YY, Luo LJ, Li Y, Ling BK, Chen BW, Wang XW, et al. A luminescent probe for highly selective Cu2+ sensing using a lanthanide-doped metal organic framework with large pores. European Journal of Inorganic Chemistry. 2019; **2019**:206-211. DOI: 10.1002/ ejic.201800945

[94] Sun Z, Sun J, Xi L, Xie J, Wang X, Ma Y, et al. Two novel lanthanide metal-organic frameworks: Selective luminescent sensing for nitrobenzene, Cu2+, and MnO4. Crystal Growth & Design. 2020;**20**:5225-5234. DOI: 10.1021/acs.cgd.0c00432

[95] Huang P, Wu F, Mao L. Target-triggered switching on and off the luminescence of lanthanide coordination polymer nanoparticles for selective and sensitive sensing of copper ions in rat brain. Analytical Chemistry. 2015;**87**: 6834-6841. DOI: 10.1021/acs. analchem.5b01155

[96] Ma Q, Zhang M, Xu X, Meng K, Yao C, Zhao Y, et al. Multiresponsive supramolecular luminescent hydrogels based on a nucleoside/lanthanide complex. ACS Applied Materials & Interfaces. 2019;**11**:47404-47412. DOI: 10.1021/acsami.9b17236

[97] He J, Wang J, Xu Q, Wu X, Dutta A, Kumar A, et al. Syntheses and crystal structures of new dinuclear lanthanide complexes based on 3-(4-Hydroxyphenyl)Propanoic acid: Hirshfeld surface analyses and photoluminescence sensing. New Journal of Chemistry. 2019;**43**:13499-13508. DOI: 10.1039/c9nj02213a

[98] Xu Q, Li Z, Li H. Water-soluble luminescent hybrid composites consisting of oligosilsesquioxanes and lanthanide complexes and their sensing ability for Cu2+. Chemistry-A European Journal. 2016;**22**:3037-3043. DOI: 10.1002/chem.201504300

[99] Wang S, Cao T, Yan H, Li Y, Lu J, Ma R, et al. Functionalization of microporous lanthanide-based metal-organic frameworks by dicarboxylate ligands with methyl-substituted thieno [2,3-b]thiophene groups: Sensing activities and magnetic properties. Inorganic Chemistry. 2016;**55**: 5139-5151. DOI: 10.1021/acs. inorgchem.5b02801

[100] Mahmoud ME, Moussa Z, Prakasam T, Li L, Abiad MG, Patra D, et al. Lanthanides based metal organic frameworks for luminescence sensing of toxic metal ions. Journal of Solid State Chemistry. 2020;**281**. DOI: 10.1016/j. jssc.2019.121031. Article ID: 121031

[101] Liu B, Huang Y, Zhu X, Hao Y, Ding Y, Wei W, et al. Smart lanthanide coordination polymer fluorescence probe for mercury(II) determination. Analytica Chimica Acta. 2016;**912**: 139-145. DOI: 10.1016/j.aca.2016. 01.044

[102] Lin J, Cheng Q, Zhou J, Lin X, Reddy RCK, Yang T, et al. Five 3D lanthanide-based coordination polymers with 3,3,6T13 topology: Structures and luminescent sensor for Hg2+ and Pb2+ Ions. Journal of Solid State Chemistry. 2019;**270**:339-345. DOI: 10.1016/j.jssc.2018.11.033

[103] Ji G, Liu J, Gao X, Sun W, Wang J, Zhao S, et al. A luminescent lanthanide MOF for selectively and ultra-high sensitively detecting Pb2+ ions in aqueous solution. Journal of Materials Chemistry A. 2017;**5**:10200-10205. DOI: 10.1039/c7ta02439h

[104] Lian C, Chen Y, Li S, Hao MY, Gao F, Yang LR. Synthesis and characterization of lanthanide-based coordination polymers for highly selective and sensitive luminescent sensor for Pb2+over mixed metal ions. Journal of Alloys and Compounds. 2017; **702**:303-308. DOI: 10.1016/j.jallcom.2017.01.260

[105] Li B, Zhou J, Bai F, Xing Y. Lanthanide–organic framework based on a 4,4-(9,9-Dimethyl-9H-Fluorene-2,7-Diyl) dibenzoic acid: Synthesis, structure and fluorescent sensing for a variety of cations and anions simultaneously. Dyes and Pigments. 2020;**172**. DOI: 10.1016/j.dyepig.2019.107862. Article ID: 107862

[106] Ding B, Liu SX, Cheng Y, Guo C, Wu XX, Guo JH, et al. Heterometallic alkaline earth-lanthanide BaII-LaIII microporous metal-organic framework as bifunctional luminescent probes of Al3+ and MnO4. Inorganic Chemistry. 2016;**55**:4391-4402. DOI: 10.1021/acs.inorgchem.6b00111

[107] Zhang PF, Yang GP, Li GP, Yang F, Liu WN, Li JY, et al. Series of water-stable lanthanide metal-organic frameworks based on carboxylic acid imidazolium chloride: Tunable luminescent emission and sensing. Inorganic Chemistry. 2019;**58**: 13969-13978. DOI: 10.1021/acs.inorgchem.9b01954

[108] Liu G, Lu YK, Ma YY, Wang XQ, Hou L, Wang YY. Syntheses of three new isostructural lanthanide coordination polymers with tunable emission colours through bimetallic doping, and their luminescence sensing properties. Dalton Transactions. 2019; **48**:13607-13613. DOI: 10.1039/c9dt02733e

[109] Mondal TK, Mondal S, Ghorai UK, Saha SK. White light emitting lanthanide based carbon quantum dots as toxic Cr (VI) and PH sensor. Journal of Colloid and Interface Science. 2019;**553**:177-185. DOI: 10.1016/j.jcis.2019.06.009

[110] Zou JY, Li L, You SY, Cui HM, Liu YW, Chen KH, et al. Sensitive luminescent probes of aniline, benzaldehyde and Cr(VI) based on a Zinc(II) metal-organic framework and its lanthanide(III) post-functionalizations. Dyes and Pigments. 2018;**159**:429-438. DOI: 10.1016/j.dyepig.2018.07.005

[111] Dang H, Li Y, Zou H, Liu S. Tunable white-light emission hybrids based on lanthanide complex functionalized poly (ionic liquid): Assembly and chemical sensing. Dyes and Pigments. 2020;**172**. DOI: 10.1016/j.dyepig.2019.107804. Article ID: 107804

[112] Chen W, Fan R, Fan J, Liu H, Sun T, Wang P, et al. Lanthanide coordination polymer-based composite films for selective and highly sensitive detection of Cr2O72- in aqueous media. Inorganic Chemistry. 2019;**58**(22):15118–15125. DOI: 10.1021/acs.inorgchem.9b01841

[113] Zhou Z, Gu J, Qiao X, Wu H, Fu H, Wang L, et al. Double protected

lanthanide fluorescence core@shell colloidal hybrid for the selective and sensitive detection of ClO. Sensors and Actuators B: Chemical. 2019;**282**: 437-442. DOI: 10.1016/j.snb.2018.11.103

[114] Bradberry SJ, Byrne JP, McCoy CP, Gunnlaugsson T. Lanthanide luminescent logic gate mimics in soft matter: [H+] and [F-] dual-input device in a polymer gel with potential for selective component release. Chemical Communications. 2015;**51**:16565-16568. DOI: 10.1039/c5cc05009j

[115] Xu W, Zhou Y, Huang D, Su M, Wang K, Xiang M, et al. Luminescent sensing profiles based on anion-responsive lanthanide(iii) quinolinecarboxylate materials: Solid-state structures, photophysical properties, and anionic species recognition. Journal of Materials Chemistry C. 2015;**3**:2003-2015. DOI: 10.1039/c4tc02369b

[116] Liu X, Yang X, Ma Y, Liu J, Shi D, Niu M, et al. Construction of two lanthanide schiff base complexes: Chiral "Triple-Decker" structure and NIR luminescent response towards anions. Journal of Luminescence. 2021;**229**. DOI: 10.1016/j.jlumin.2020.117679. Article ID: 117679

[117] Sun Z, Hu P, Ma Y, Li L. Lanthanide organic frameworks for luminescence sensing of nitrobenzene and nitrophenol with high selectivity. Dyes and Pigments. 2017;**143**:10-17. DOI: 10.1016/j.dyepig.2017.04.015

[118] Khullar S, Singh S, Das P, Mandal SK. Luminescent lanthanide-based probes for the detection of nitroaromatic compounds in water. ACS Omega. 2019;**4**:5283-5292. DOI: 10.1021/acsomega.9b00223

[119] Zhao SN, Song XZ, Zhu M, Meng X, Wu LL, Song SY, et al. Highly thermostable lanthanide metal-organic frameworks exhibiting unique selectivity for nitro explosives. RSC Advances. 2015;**5**:93-98. DOI: 10.1039/c4ra13773f

[120] Wang X, Zhang L, Yang J, Liu F, Dai F, Wang R, et al. Lanthanide metal-organic frameworks containing a novel flexible ligand for luminescence sensing of small organic molecules and selective adsorption. Journal of Materials Chemistry A. 2015;**3**:12777-12785. DOI: 10.1039/c5ta00061k

[121] Yang LZ, Wang J, Kirillov AM, Dou W, Xu C, Fang R, et al. 2D lanthanide MOFs driven by a Rigid 3,5-Bis(3-Carboxy-Phenyl)pyridine building block: Solvothermal syntheses, structural features, and photoluminescence and sensing properties. CrystEngComm. 2016;**18**:6425-6436. DOI: 10.1039/c6ce00885b

[122] Einkauf JD, Ortega RE, Mathivathanan L, De Lill DT. Nitroaromatic sensing with a new lanthanide coordination polymer [Er2 (C10H4O4S2)3(H2O)6]:N assembled by 2,2′-Bithiophene-5,5′-Dicarboxylate. New Journal of Chemistry. 2017;**41**: 10929-10934. DOI: 10.1039/c7nj01677h

[123] Liu W, Huang X, Chen C, Xu C, Ma J, Yang L, et al. Function-oriented: The construction of lanthanide MOF luminescent sensors containing dual-function urea hydrogen-bond sites for efficient detection of picric acid. Chemistry—A European Journal. 2019; **25**:1090-1097. DOI: 10.1002/chem.201805080

[124] Wang JM, Zhang PF, Cheng JG, Wang Y, Ma LL, Yang GP, et al. Luminescence tuning and sensing properties of stable 2D lanthanide metal-organic frameworks built with

symmetrical flexible tricarboxylic acid ligands containing ether oxygen bonds. CrystEngComm. 2021;**23**:411-418. DOI: 10.1039/d0ce01528h

[125] Moscoso FG, Almeida J, Sousaraei A, Lopes-Costa T, Silva AMG, Cabanillas-Gonzalez J, et al. A lanthanide MOF immobilized in PMMA transparent films as a selective fluorescence sensor for nitroaromatic explosive vapours. Journal of Materials Chemistry C. 2020; **8**:3626-3630. DOI: 10.1039/d0tc00376j

[126] Fu R, Hu S, Wu X. Rapid and sensitive detection of nitroaromatic explosives by using new 3D lanthanide phosphonates. Journal of Materials Chemistry A. 2017;**5**:1952-1956. DOI: 10.1039/c6ta10152f

[127] Li XZ, Zhou LP, Yan LL, Yuan DQ, Lin CS, Sun QF. Evolution of luminescent supramolecular lanthanide M2nL3n complexes from helicates and tetrahedra to cubes. Journal of the American Chemical Society. 2017; **139**:8237-8244. DOI: 10.1021/jacs.7b02764

[128] Zhang F, Chen S, Nie S, Luo J, Lin S, Wang Y, et al. Waste PET as a Reactant for Lanthanide MOF Synthesis and Application in Sensing of Picric Acid. Polymers. 2019;**11**:2015. DOI: 10.3390/polym11122015

[129] Raizada M, Sama F, Ashafaq M, Shahid M, Khalid M, Ahmad M, et al. Synthesis, structure and magnetic studies of lanthanide metal–organic frameworks (Ln–MOFs): Aqueous phase highly selective sensors for picric acid as well as the arsenic ion. Polyhedron. 2018; **139**:131-141. DOI: 10.1016/j.poly.2017.09.052

[130] Wang X, Yan P, Li Y, An G, Yao X, Li G. Highly efficient white-light emission and UV-Visible/NIR luminescence sensing of lanthanide metal-organic frameworks. Crystal Growth & Design. 2017;**17**: 2178-2185. DOI: 10.1021/acs.cgd.7b00112

[131] Shi Y, Wang WM, Tang GP, Zhang YX, Li M, Wu ZL. Robust lanthanide metal–organic frameworks with highly sensitive sensing of aniline and slow magnetization relaxation behaviors. Polyhedron. 2018;**153**:122-127. DOI: 10.1016/j.poly.2018.07.003

[132] Shi B, Zhong Y, Guo L, Li G. Two dimethylphenyl imidazole dicarboxylate-based lanthanide metal-organic frameworks for luminescence sensing of benzaldehyde. Dalton Transactions. 2015;**44**:4362-4369. DOI: 10.1039/c4dt03326d

[133] Han Y, Yan P, Sun J, An G, Yao X, Li Y, et al. Luminescence and white-light emitting luminescent sensor of tetrafluoroterephthalate-lanthanide metal-organic frameworks. Dalton Transactions. 2017;**46**:4642-4653. DOI: 10.1039/c7dt00215g

[134] Yao X, Wang X, Han Y, Yan P, Li Y, Li G. Structure, color-tunable luminescence, and UV-Vis/NIR benzaldehyde detection of lanthanide coordination polymers based on two fluorinated ligands. CrystEngComm. 2018;**20**:3335-3343. DOI: 10.1039/c8ce00516h

[135] Wang L, Fan G, Xu X, Chen D, Wang L, Shi W, et al. Detection of polychlorinated benzenes (persistent organic pollutants) by a luminescent sensor based on a lanthanide metal-organic framework. Journal of Materials Chemistry A. 2017;**5**:5541-5549. DOI: 10.1039/c7ta00256d

[136] Zhu Y, Wang L, Chen X, Wang P, Fan Y, Zhang P. 3D lanthanide metal-organic frameworks constructed from 2,6-naphthalenedicarboxylate ligand: Synthesis, structure, luminescence and dye adsorption. Journal of Solid State Chemistry. 2017;**251**:248-254. DOI: 10.1016/j.jssc.2017.04.026

[137] Du N, Gao X, Song J, Wang ZN, Xing YH, Bai FY, et al. Optical detection of small biomolecule thiamines at a micromolar level by highly luminescent lanthanide complexes with tridentate N-heterocyclic ligands. RSC Advances. 2016;**6**:71012-71024. DOI: 10.1039/c6ra10869e

[138] Wang Y, Du N, Zhang X, Wang Y, Xing YH, Bai FY, et al. Functional sensing materials based on lanthanide N-heterocyclic polycarboxylate crystal frameworks for detecting thiamines. Crystal Growth & Design. 2018;**18**: 2259-2269. DOI: 10.1021/acs.cgd.7b01684

[139] Lian X, Yan B. A lanthanide metal-organic framework (MOF-76) for adsorbing dyes and fluorescence detecting aromatic pollutants. RSC Advances. 2016;**6**:11570-11576. DOI: 10.1039/c5ra23681a

[140] Zhang WW, Wang YL, Liu Q, Liu QY. Lanthanide-benzophenone-3,3′-Disulfonyl-4,4′-dicarboxylate frameworks: Temperature and 1-hydroxypyren luminescence sensing and proton conduction. Inorganic Chemistry. 2018;**57**:7805-7814. DOI: 10.1021/acs.inorgchem.8b00865

[141] Singha DK, Majee P, Mondal SK, Mahata P. Detection of pesticide using the large stokes shift of luminescence of a mixed lanthanide co-doped metal–organic framework. Polyhedron. 2019; **158**:277-282. DOI: 10.1016/j.poly.2018.10.066

[142] Ren K, Wu SH, Guo XF, Wang H. Lanthanide organic framework as a reversible luminescent sensor for sulfamethazine antibiotics. Inorganic Chemistry. 2019;**58**:4223-4229. DOI: 10.1021/acs.inorgchem.8b03284

[143] Zhang F, Yao H, Chu T, Zhang G, Wang Y, Yang Y. A lanthanide MOF thin-film fixed with Co3O4 nano-anchors as a highly efficient luminescent sensor for nitrofuran antibiotics. Chemistry—A European Journal. 2017; **23**:10293-10300. DOI: 10.1002/chem.201701852

[144] Wu S, Zhu M, Zhang Y, Kosinova M, Fedin VP, Gao E. Luminescent sensors based on coordination polymers with adjustable emissions for detecting biomarker of pollutant ethylbenzene and styrene. Applied Organometallic Chemistry. 2021;**35**:e6058. DOI: 10.1002/aoc.6058

[145] Belousov YA, Drozdov AA, Taydakov IV, Marchetti F, Pettinari R, Pettinari C. Lanthanide azolecarboxylate compounds: Structure, luminescent properties and applications. Coordination Chemistry Reviews. 2021; **445**:214084. DOI: 10.1016/j.ccr.2021.214084

Chapter 2

Perspective Chapter: History and Classification of Gold Mineralization in Egypt

Refaey El-Wardany and Jiangang Jiao

Abstract

Gold deposits and occurrences are spread over a large number of sites which covered by Precambrian the basement rocks in the Eastern Desert of Egypt, and the gold exploration and mining activities had been districted since ancient times. The ancient Egyptians extracted gold from quartz veins of various dimensions in open pits and underground workings. Consequently, gold production in Egypt seems to have started as early as the predynastic times (about 6000 BCE) from Pre-and Early Dynastic times to Early Arab Period. In the modern days, Egyptian mineral resource authority announced many big international bid-rounds for Gold Exploration and associated minerals was launched to open the minerals sector to foreign investment in the mining industry. Many studies have been achieved to classify the gold deposits in Egypt based on various characteristics such as the nature, occurrences, type of host rock, tectonic setting, metal association, fluid inclusion, and the tectonic environment of mineralization. Gold deposit in Egypt still need more geological and mining studies to understand the gold system types, additionally, to get answers regarding to source, transportation, traps, and host rocks, as well as the whole challenges in mining sectors.

Keywords: gold deposit, Eastern Desert, gold exploration, classification of gold, ancient Egyptians

1. Introduction

The Arabian-Nubian Shield (ANS) is one of the largest and best-recorded outcrops of Neoproterozoic Juvenile rocks on the Earth (950–450 Ma), the ANS developed during the collision between East and West Gondwana with the closure of the Mozambique Ocean (Pacific-sized Ocean) at the north part of the East African orogen (EAO) (**Figure 1a** and **b**). The East African Orogen is a part of the Pan-African orogenic cycle that extends from the north by ANS to the south of the Mozambique belt. The Pan-African Orogeny was a period of major crustal accretion and a part of the Neoproterozoic Gondwana supercontinent; it affected Gondwana and several regions in Laurasia [2, 5–9]. The Egyptian Nubian Shield (ENS), the basement rocks of Egypt comprise covers 100,000 km^2 (about 10% of the total area of the country), exposed mainly in the southern Sinai Peninsula, the Eastern Desert parallel to the eastern flank of the Red Sea coast, and the Western Desert at Gabal Oweinat. Commonly, the Eastern Desert are divided into three distinct lithotectonic

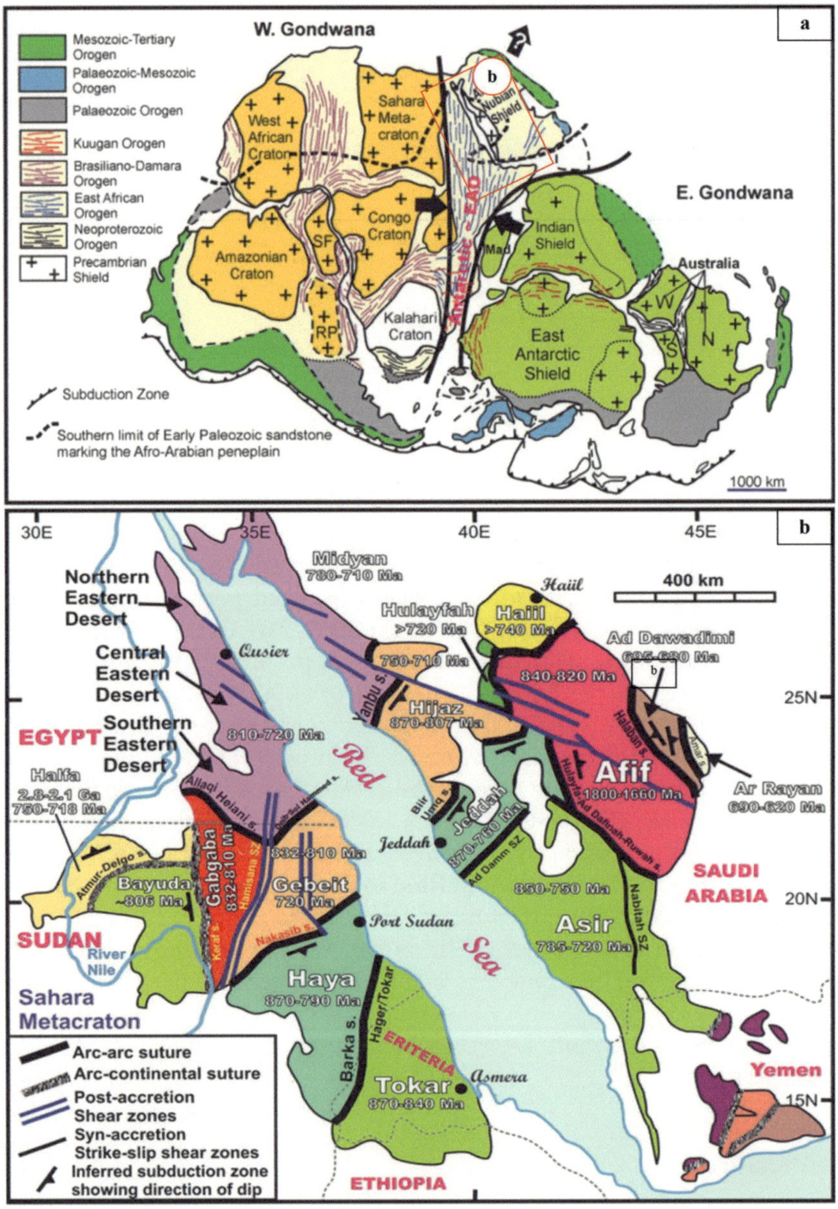

Figure 1.
(a) Showing the Gondwanaland supercontinent after [1], (b) Showing locations, terranes, and sutures of Arabian–Nubian shield after [1–4].

domains as specific: Northern Eastern Desert (NED), Central Eastern Desert (CED), and the Southern Eastern Desert (SED) (**Figure 2a**) [11–15]. The basement complex in the Eastern Desert is composed of a heterogeneous rock assemblage; it can be classified into two major units (**Figure 2b**). The lower tectonostratigraphic unit (infrastructural unit) of the old continental margin comprises medium to high-grade metamorphic gneisses and migmatites, psammitic schists (e.g., Meatiq, Migif-Hafafit, Sibai, and El-Shalul domes) [16–19]. The upper tectonostratigraphic unit (suprastructural unit) [13, 18–21], which is represented by ophiolites, magmatic-arc related island-arc volcanic, and volcaniclastic

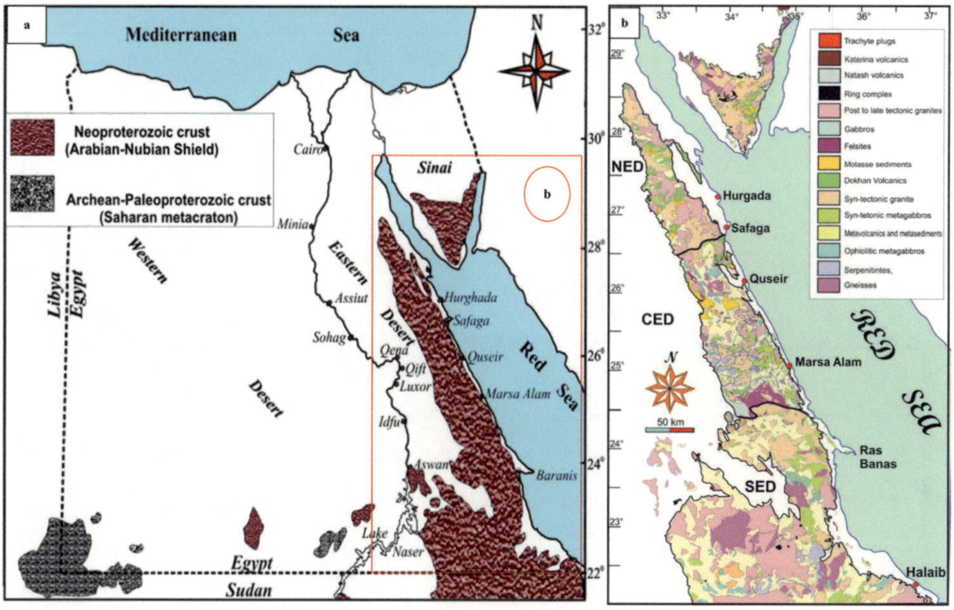

Figure 2.
(a) Distribution of Neoproterozoic basement rocks in Egypt [10], (b) Geologic map of the Eastern Desert and Sinai [1].

intruded by Syn to late tectonic calc-alkaline and alkaline granites, metagabbro-diorite complexes, Dokhan volcanics, and molasse sediments intruded by post tectonic granitic intrusions and dike [10, 13, 22–28].

2. History of gold mining in Egypt

The history of gold mining in Egypt can be divided into two main times: ancient time and modern time.

2.1 History of ancient gold mining

The production and mining activities of gold deposits have been carried out from the Eastern Desert of Egypt since ancient times, more than 240 gold deposits and occurrences are spread over the whole area covered by the basement rocks of the Precambrian age.

In most mines, the ancient Egyptians extracted gold from quartz veins of various dimensions in open pits and underground workings. Gold production in Egypt seems to have started as early as the predynastic times (about 4000 BC) and continued in different periods up to the sixth century when the gold mines lost their importance [29–35]. Mond and Winkler published a preliminary report in the season 1937–1938; they noted that the historical time of gold production was started with the accumulation of the nugget gold from the grounds of some wadis in the middle of 5000 BC, which dated by the evidence in production sites of by the "Earliest Hunters" [36].

Harrell, 1992, submitted an article about the Turin papyrus map one of the oldest maps and the oldest map that illustrates the topography and geology of Wadi Hammamat in the Central Eastern Desert of Egypt (**Figure 3**). However,

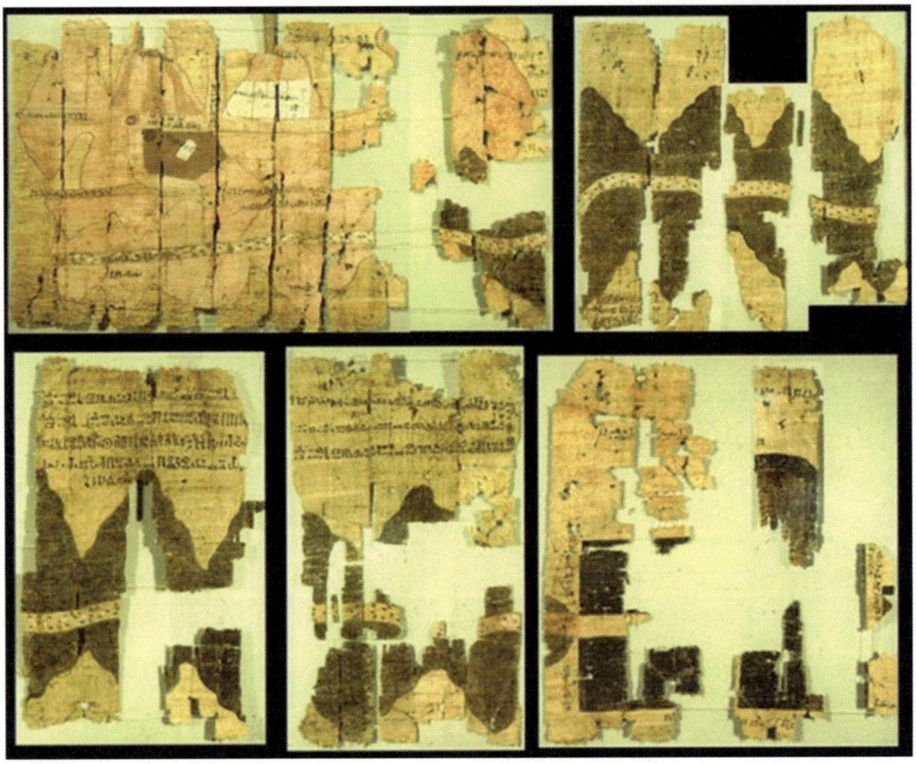

Figure 3.
The Turin papyrus map illustrates the topography, geology and gold working at Wadi Hammamat during the reign of Ramesses IV (1151–1145 B.C.) [37].

it shows gold-working settlements during the ancient time at Bir Umm Fawakhir and "bekhen-stone" quarry regions. He noted that the map was drawn by the chief administrative officer ("Scribe of the Tomb" Amennakhte, son of Ipuy) of the village of Deir El-Medina during the reign of Ramesses IV (1151–1145 BC) to record the king's bekhen-stone quarrying expedition to Wadi Hammamat late in the third year of his reign [37]. Generally, the gold production periods in Egypt can be classified into eight periods: Pre-and Early Dynastic times (~3500–2700 BC), Old and Middle Kingdom (2700–2160 and 2119–1794 BC, respectively), New Kingdom times to the Third Intermediate Period (1550–1070 BC), Ptolemaic (Greek) times (~300–30 BC), Kingdoms of Kushitic times (800–400 BC and to 100 AD), Roman and Byzantine Period times (~30 BC–641 AD), Early Arab Period (~990–1350 AD), and modern mining time. Each period includes certain techniques and equipment [26, 34–39].

2.2 Modern gold mining history

In the nineteenth century, Egyptian gold attracted increasing attention, and at the beginning of the present century, most of the gold deposits were rediscovered and worked. However, by the end of 1918, gold mining practically stopped [29, 40]. The period between 1932 and 1958 represents a new stage of gold exploration in Egypt, during which several deposits were examined and the major ones, such as Sukari, Um Ud, Hangaliya, Um Rus, Barramiya, El Sid, Um Garaiyat, and others, were put under exploitation. From 1902 to 1958, the amount of pure gold recovered in Egypt was about 7 tons [1, 40]. The last stage in the history of gold mining is the prospecting work that was carried out at Atud deposits in 1953–1954 [41] and 1963–1964 [42]. In 1967, the gold deposits of Semna, Atalla, El Sid, Um Rus, Barramiya, Atud, Sukari, Um Ud, Hangaliya, Kurdeman, Hamish, Um Garaiyat, and others were examined by [43]. Gold deposits and occurrences in Egypt have been reviewed in many publications, some data on the gold deposits were presented in the reports of the Mining Guidance, 1935–1958, reports of Mines and Quarries Department 1906–1960, reports of the Egyptian Mining and prospecting Co., 1955–1962 and Hunting Ltd., 1967. There are many different views concerning the stages of gold development and the relations of gold mineralization with various magmatic complexes. In the period from 1983 to 1989 a new stage in the history of gold mining when the work was carried out at many gold mines in the Eastern Desert by Minex Company. The Geological Survey of Egypt played an important role in the prospecting work done by Minex Company and gave the most accurate and complete data on many gold mines, for example, Abu Marrawat, El Barramiya, HI Fawkhir, and Semena.

In 1990, the Pharaonic Company begins to prospect gold in Egypt depending on modern theories and using more accurate techniques for analysis, depending on data collected by Minex. Nowadays, Cryst International Company seeks gold in the Eastern Desert depending on modern theories. Cryst International Company is interested in gold associated with pyrite in many locations, especially in the Hammata area. In 1995, many companies invested in gold exploration in Egypt (i.e., Centamin, Aton Resources Inc. (formerly Alexander Nubia), Thani Dubai, Mica Star, SMW,

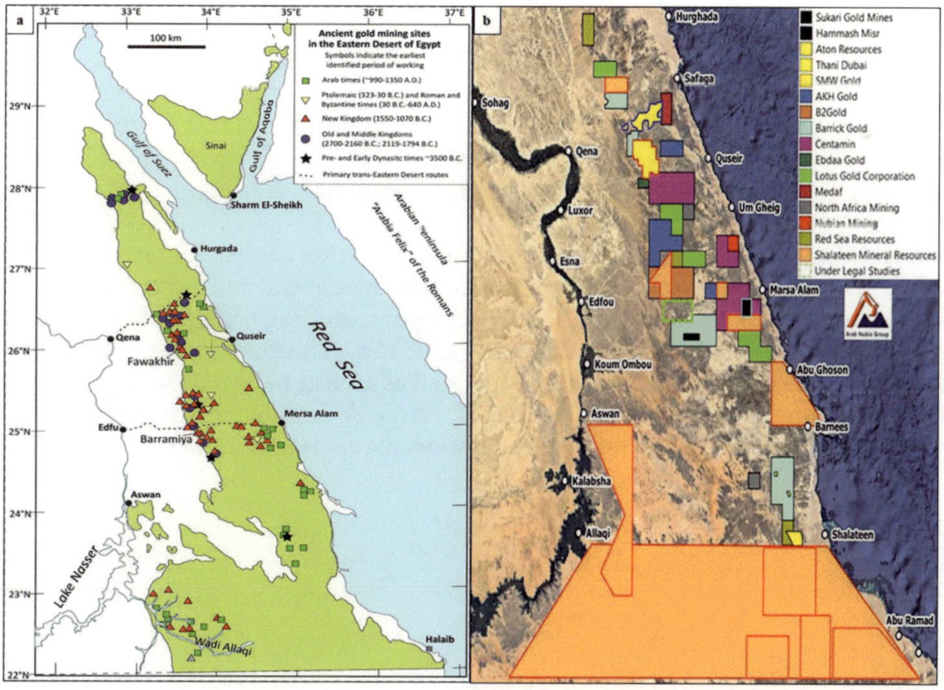

Figure 4.
(a) Map shows the historical regions of gold in the Eastern Desert according to the earliest known ancient periods working [40], and (b) concession of the international and Egyptian companies in 2020 (after Arab Nubian group https://www.linkedin.com/company/5013242/admin/).

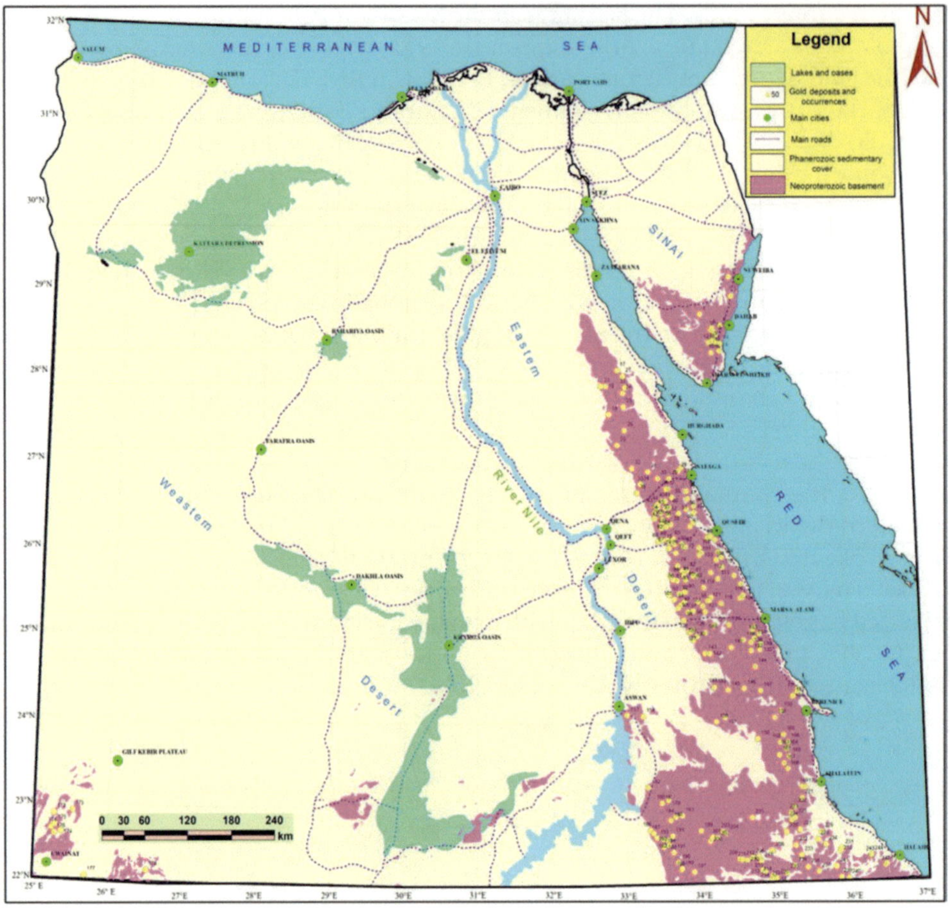

Figure 5.
Location map of gold deposits and occurrences in Eastern Desert, Egypt [44, 45].

Matz Holding, and others). Although all these exploration companies targeted well-known occurrences of gold mineralization in the Central and Southern Eastern Desert, only the Sukari gold mine, in the southern part of the CED, has been successfully developed and mined by the Centamin Company. The Sukari deposit is the only producing mine in Egypt, and now it is operated as a shear between the Egyptian Government and Centamin plc. It consists of a large-scale open pit and a high-grade underground mine. The total mineral reserves were reported at 7.25 Moz and the total

1	Nuweiba	70	Wadi Mahasin	139	Um Ud	208	E-G.Um El Tuyour El Fuqani
2	Nisriyin	71	Wadi Badah al atshani	140	Hangaliya	209	Um El Tuyour
3	A1 Reigata	72	Wadi Zaraib	141	Um Huqab	210	W-G.Um El Tuyour El Fuqani
4	Dahab	73	Hamouda	142	Doweig	211	Gabal Al Adraq
5	South Gabal Ferani	74	Abu Fannani	143	Wadi Um Hagalic	212	Betan – II

6	Wadi Nasb Area	75	Um Soleimat	144	Hamash	213	Betam
7	South Wadi El Shalal	76	El Sid	145	Geili	214	El Hoteib
8	Abu Rahab	77	El Shihimiya Um Hargal	146	Khashab	215	Um Egat – 1
9	Wadi Madsus	78	El Nur	147	North Gabal El Khellah	216	Um Egat – II
10	Um Zariq – I	79	Zeidon	148	Gabal Abu Himimid	217	Egat Mine
11	South Wadi Nasb	80	Makhyt	149	Kab El Rayan	218	Shanaiyet
12	Um Zariq – II	81	Shughief	150	El Qulan	219	Al Fawi –1
13	Wadi El Kid	82	Wadi Zeidon	151	Wadi El Homer	220	Al Fawi – II
14	Wadi Qabila	83	Hadarba	152	Shialek	221	Al Fawi – III
15	Wadi El Samra	84	WadiUmAash	153	Abu Rahaya	222	Wadi Beida
16	Wadi Um Toleihat	85	Maiser	154	Kortonos	223	Wadi El Abied
17	Gabal Dara	86	Perevalny	155	El Hudy	224	Korbiai
18	Wadi Zobeir	87	Wadi Manih	156	Wadi I Hi Mikan	225	Um Radam
19	Um Balad	88	El Heir	157	Zergat Naam	226	North Gabal Gerf
20	Ruh A1 Hadid	89	Tigelny (Wadi El Heir)	158	Gabal Nikeiba	227	El Direib
21	Wadi Dara	90	Wadi El Moilh	159	Urf Abu Humar	228	Madarai
22	Um Mongul	91	Daghbag	160	Abu Gurdi	229	Wadi Oregium
23	Wadi El Urf	92	Deghig	161	Salib El Abied	230	El Qurun
24	Wadi Dib	93	Tila Gadalla	162	Betan – 1	231	Romeit
25	Talat A1 Zarqa	94	Fella Godalla	163	Um Eleiga	232	Groore
26	Um Tweir	95	El Hisinat	164	Ourga El Rayan – 1	233	Abu Hireig
27	El Helliev	96	El Mayia (El Bakriya)	165	Orgarian	234	Gabal Abu Hodied
28	Abu Morrat	97	Um Salatite	166	Hutit	235	W. Meisah
29	Um Tagher	98	Um Selim	167	Um Kalib	236	Gabal Heianai –1
30	Wadi Fatira El Zarqa	99	Al Barramiya	168	Ourga El Rayan – II	237	Heianai – II
31	Fatiri El Beida	100	Bokari-I	169	Wadi Khuda	238	Gabal Suruk
32	El Ghozah	101	Bokari-II	170	El Anbat	239	Soaorib (Gudir)
33	Fatiri	102	Abu Mouawad	171	Southwest Gabal Sirsir	240	Wadi El Qurat
34	Wadi El Dob	103	Kap El Abiad	172	Gabal Nazar area-I	241	Gabal Shiab

35	Abu Shehat	104	Tarfawi	173	Gabal Nazar area-I I	242	Gabal Hamida
36	El Bolh	105	Wadi Kareim	174	Gabal Nazar area-I 11	243	Gabal Elba
37	Gray	106	Wadi Kareim-BIF	175	Gabal Nazar area-VI	244	Sul Hamid
38	Safaga	107	El Dabbah	176	North Gabal Peter and Paul	245	Qash Amir
39	Abu Marawat	108	Hamrat ghannam	177	Southwest Gabal Nazar	246	Oseir Irab
40	Semna	109	Sharm El Bahari	178	Gabal Kamel		
41	Massaghat El Gukh	110	Abu mureiwa	179	South Gabal Kamel		
42	Gabal Semna	111	Nusla	180	Hisayrbah		
43	E 1 Eredeia	112	Um Grifate	181	Hariari		
44	Bohlog	113	Atwi	182	Um Araka		
45	Gidami	114	Um Ghamis	183	Um Ashirah		
46	Abu Qarahish	115	Sigdit	184	Nekib		
47	Wadi Gasus	116	UmRus	185	Abu Swayel		
48	Wadi Abu Shigeili	117	WadiMurin	186	Middle Wadi Haimur		
49	Wadi Queih	118	AbuDabbab	187	Haimur		
50	Wadi Queih	119	UmNar	188	Wadi El Nile		
51	Sagi	120	Gabal El Hadid	189	Atshan		
52	Wadi Hemeiriya	121	Um Samra-I	190	Murra		
53	Kap Amiri	122	Um Samra-II	191	Korsco		
54	A1 Areifia	123	Abu Quraiya	192	Atshani		
55	Hamama-I	124	Beizah	193	Maraheib		
56	Hamama-II	125	Atud 193	194	Um Garaiyat		
57	Abu Had	126	UrfElFahd	195	South Kalabsha		
58	Atalla El Mor	127	Abu Mireiwa	196	Fillat		
59	Kap El Abse	128	Dungash West	197	Um Doma		
60	Rabshi	129	Dungash East	198	Abu Fass		
61	Wadi Sodmein	130	Um Dalalil	199	Eneigit – 1		
62	El Haramiya	131	Samut	200	Eneigit – II		
63	North Wadi Atalla	132	Um Tundeba	201	Seiga – 1		
64	Wadi Nabsh El Qadim	133	Sukari	202	Seiga – II		
65	Atalla	134	Am Baut	203	Um Shashoba		

66	A1 Fawakheir	135	Sabahia	204	Gabal Kulyeit – 1
67	Hammamat – I	136	El Lawi	205	Gabal Kulyeit – II
68	Hammamat – II	137	Leweiwi (Um Gamil)	206	South Mashbih
69	Um Had	138	Kurdman	207	Gabal Anweib

Table 1.
Locations of gold deposits and occurrences in Eastern Desert, Egypt.

resources at 11 Moz with a lifetime of 15 years [38–41]. The gold production started from the open pit mine in 2010 and underground started from 2011 to nowadays. The random mining or illegal gold rushers phenomena (they are called Dahaba) have been widely distributed along the whole Eastern Desert. Even though they represented some advantages in gold exploration on different scales but they display enormous harmful impacts by destroying the old mine and having a great environmental impact; at the first, they used a metal detector in exploration, but now they developed their techniques by using crushers, panning, and grinding; in addition, they used mercury and even cyanide to extract and refractory gold. In the Southern Eastern Desert, Shalateen Mineral Resources Company (SMRC), a governmental company is working to control these phenomena by giving licenses to the rushers to explore legally. In 2018, new recommendations for policies were announced by the Egyptian government to encourage global investment in mining. In August 2019, the law of mineral resources was modified and published in January 2020. In 2020, a big international bid-round No (1) for gold exploration and associated minerals was launched to open the minerals sector to foreign investment in the mining industry. At the end of 2020, EMRA announced that 11 international and Egyptian companies)Hammash Misr, Aton Resources, Thani Dubai, SMW Gold, AKH Gold, B2Gold, Barrick Gold, Centamin, Ebdaa Gold, Lotus Gold Corporation, Mcdaf, North Africa Mining, Nubian Mining, Red Sea Resources, and Shalateen Mineral Resources) win 82 gold exploration areas in Egypt (**Figure 4**). More than 195 gold deposits and occurrences are situated in the Eastern Desert of Egypt since ancient times until modern days (**Figure 5** and **Table 1**).

3. Classification of the gold deposits in Egypt

Many studies classified the gold deposits in Egypt based on different characteristics (e.g., the nature, occurrences, type of host rock, tectonic setting, metal association, and the tectonic environment of mineralization). Hume [29] believed that the formation of gold took place in two different stages and related the main gold mineralization to the second stage of hydrothermal activity that accompanied the development of the dioritic intrusion of the Metarchean age (i.e., Proterozoic) [33]. He grouped the mineral deposits into four categories: (1) gold occurrences; (2) group occurrences of silver, copper, zinc, molybdenum, tungsten, iron, chromium, nickel, lead, tin, platinum, and graphite; (3) precious and semiprecious minerals; and (4) ornamental stones. Kochine and Bassyuni [46] classified the gold mineralization depending on the mode of occurrences and the nature of mineralization into three distinguished types of gold dykes, veins, and placer deposits. El Ramly et al. [47] added that most gold deposits are confined either to the intrusive masses of granodiorites and diorites or to the schists in the close vicinity of these masses. On the other hand, he stated that there are more than 95 locations of different gold deposits and occurrences that are hosted by variable rock composition including, schists, mudstone-greywacke-conglomerates, granites, granodiorite (gabbros, metavolcanics, and dyke like felsite porphyries and trachytes, diorites). Sabet et al. [48] identified four main epochs based on the time of mineralization and formation of gold: Preorogenic epoch, Syn to late-orogenic epoch, Riphean-lower Paleozoic epoch, and Mesozoic-Cenozoic epoch. Also classified the major gold "Formations" in the Eastern Desert into: (1) gold-sulfide "Formation", (2) gold-bearing iron quartzite "Formation", and (3) gold-bearing quartz veins "Formation". Sabet and Bordonosov [49] classified the occurrences of gold in Egypt into three types, namely, gold-sulfide, skarn gold-ferruginous quartzite, and gold quartz veins formations. The gold quartz vein formation was subdivided according to the metal and mineral associations into: gold-arsenic, gold-pyrite, gold-polymetal, gold-copper, gold-mercury, and

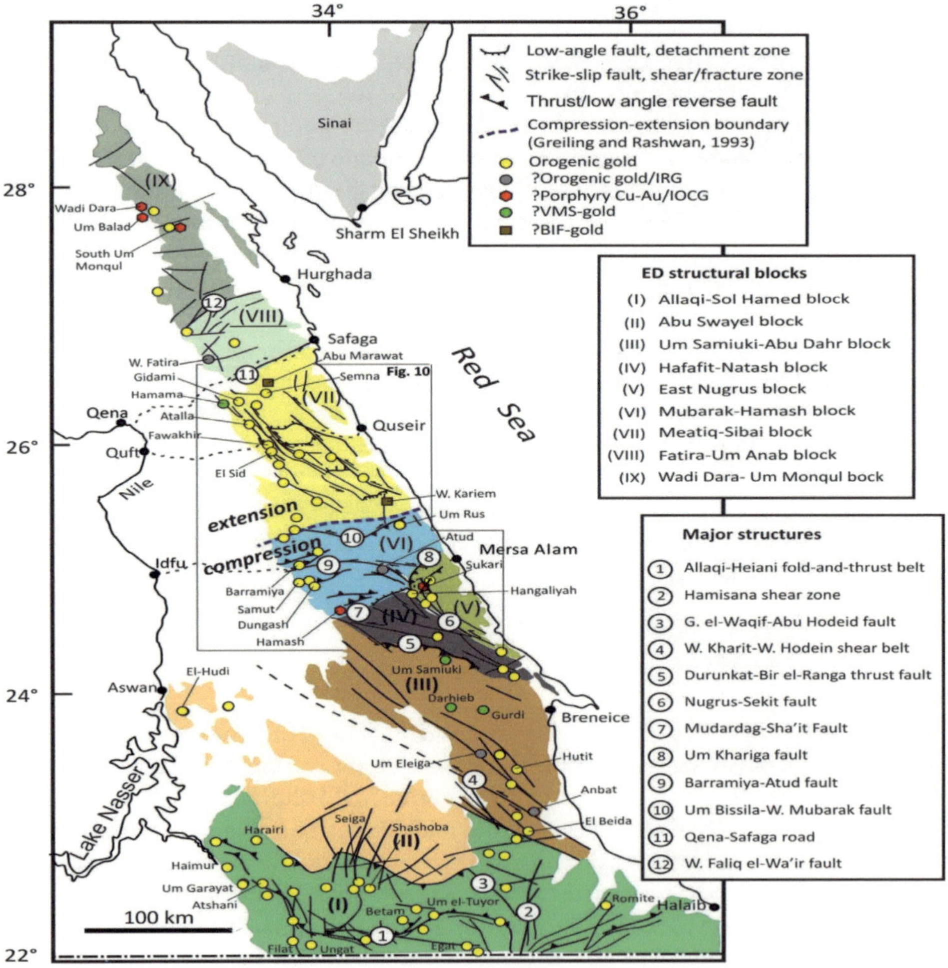

Figure 6.
Gold occurrences and structural blocks in ED [39].

gold-antimony types. Botros [50] classified the gold mineralization in the Eastern Desert depending on their related tectonic environment into three main classes and subdivided these classes depending mainly on the type of the host rock as follows: gold hosted in the island-arc stage (such as BIF and tuffaceous metasedimentary rocks), gold hosted in the orogenic stage (such as gold mineralization along the sheared contacts of the ophiolitic serpentinites, associated with porphyry copper mineralization, and auriferous quartz veins intruded gabbros and granodiorite, and the third type is gold mineralization going with the post-orogenic stage. Botros [37], according to the tectonic-magmatic evolution of the Nubian shield, modified his classification of gold deposits to new threefold gold deposits. (1) Stratabound deposits are subdivided into three main types: gold-bearing Algoma-type banded iron formation, gold-bearing tuffaceous sediments, and gold-bearing volcanogenic massive sulfide deposits; (2) non-stratabound deposits are divided into two main types: vein-type mineralization hosted in a wide range of rocks and disseminated-type mineralization hosted in hydrothermally altered rocks (alteration zones); and (3) placer deposits are divided into modern placers and lithified placers. Zoheir et al. [39] noted that the Eastern Desert is divided into nine structure blocks, which are the Allaqi-Sol Hamed block, Abu Swayel block, Um Samiuki-Abu Dahr block, Hafafit-Natash block, East Nugrus block, Mubarak-Hamash block, Meatiq-Sibai block, Fatira-Um Anab block, and Wadi Dara-Um Monqul bock (**Figure 6**). They reviewed that the gold deposits in Egypt are mainly orogenic gold systems, but some other gold systems with some dough and limited distribution have been studied as volcanic massive sulfide (VMS), porphyry systems, Fe-oxide-Cu-Au (IOCG) systems, and intrusion-related system. On the other hand, they mentioned that the gold deposits are controlled by structure and just five that controlled the orogenic gold occurrences are Allaqi-Sol, Hamed block, Um Samiuki-Abu Dahr block, East Nugrus block, Mubarak-Hamash block, and Meatiq-Sibai block. El Aref et al. [51] believed that most of the gold deposits in the Eastern Desert are related to orogenic gold deposits and are structurally controlled. They noted that there is some evidence of VMS system, associated high-sulfidation Au-Cu-Mo mineralization and porphyry type, while they mentioned that there is no evidence in support of the presence of iron oxide copper-gold (IOCG) or even intrusion-related in the Eastern Desert.

4. Conclusion

The Arabian-Nubian Shield (ANS) is one of the largest and best-recorded outcrops of Neoproterozoic Juvenile rocks on the Earth; the basement rocks extend in the Eastern Desert are divided into three distinct lithotectonic domains as specific: Northern Eastern Desert (NED), the Central Eastern Desert (CED), and the Southern Eastern Desert (SED). The basement complex in the Eastern Desert is composed of a heterogeneous rock assemblage; it can be classified into the lower tectonostratigraphic unit or infrastructural and the upper tectonostratigraphic unit or suprastructural. The production and mining activities of gold deposits have been conducted in the Eastern Desert of Egypt since ancient times; nowadays, the random mining or illegal gold rushers phenomena (they are called Dahaba) have been widely distributed along the whole Eastern Desert. Many studies classified the gold deposits in Egypt based on distinctive characteristics, such as nature, occurrences, type of host rock, tectonic setting, metal association, and the tectonic environment of mineralization. Still, there are many challenges in the geologic exploration and mining studies of gold in Egypt.

Acknowledgements

The authors would like to thank the National Science Foundation of China (NSFC: 92162213). They express their gratitude to the group of School of Earth Science and Resources, Chang'an University for their support during this research. Refaey El-Wardany is very grateful to Prof. Ibrahem Abu El-Leil, Prof. Basem Zohair, Prof. Yonggang Feng, Dr. Amr Abdelnasser, in addition, the Geology Department Faculty of Science Al-Azhar University (Assiut Branch) and the China Scholarship Council, for help and supporting his research. The editor and the anonymous reviewers are acknowledged for their many insightful comments and suggestions that improved many interpretations.

Author details

Refaey El-Wardany[1,2] and Jiangang Jiao[1*]

1 School of Earth Science and Resource, Chang'an University, Xi'an, Shaanxi, China

2 Department of Geology, Faculty of Science, Al-Azhar University, Assiut Branch, Egypt

*Address all correspondence to: jiangang@chd.edu.cn-64980216@qq.com

References

[1] Abd El-Wahed M, Hamimi Z. The Egyptian Nubian Shield Within the Frame of the Arabian–Nubian Shield. In: Hamimi Z, Arai S, Fowler AR, El-Bialy MZ, editors. The Geology of the Egyptian Nubian Shield. Cham: Springer; 2021. DOI: 10.1007/978-3-030-49771-2_2

[2] Johnson PR, Woldehaimanot B. Development of the Arabian-Nubian Shield. Perspectives on accretion and deformation in the northern East African Orogen and the assembly of Gondwana. In: Yoshida M, Windley BF, Dasgupta S, editors. Proterozoic East Gondwana, Supercontinent Assembly and Breakup. London: Geological Society; 2003. pp. 290-325

[3] Hargrove US, Stern RJ, Kimura J-I, Manton WI, Johnson PR. How juvenile is the Arabian–Nubian Shield? Evidence from Nd isotopes and pre-Neoproterozoic inherited zircon in the Bi'r Umq suture zone, Saudi Arabia. Earth Planet Science Letters. 2006;**252**(34):308-326

[4] Abdelsalam MG. Quantifying 3D post-accretionary tectonic strain in the Arabian-Nubian Shield: superimposition of the Oko Shear Zone on the Nakasib Suture, Red Sea Hills, Sudan. Journal of African Earth Sciences. 2010;**56**(4-5):167-178. DOI: 10.1016/j.jafrearsci.2009.07.003

[5] Johnson P, Abdelsalam M, Stern B. The Bi'r Umq-Nakasib Suture Zone in the Arabian-Nubian Shield: A Key to Understanding Crustal Growth in the East African Orogen. Gondwana Research. 2003;**6**:523-530. DOI: 10.1016/S1342-937X(05)71003-0

[6] Stern RJ, Johnson PR, Kroner A, Yibas B. Neoproterozoic ophiolites of the Arabian-Nubian Shield. Developments in Precambrian Geology. 2004;**13**:95-128

[7] Ali KA, Stern RJ, Manton WI, Johnson PR, Mukherjee SK. Neoproterozoic diamictite in the Eastern Desert of Egypt and northern Saudi Arabia: evidence of *750 Ma glaciation in the Arabian-Nubian Shield? International Journal of Earth Sciences. 2010;**99**:705-726

[8] Fritz H, Abdelsalam M, Ali K, Bingen B, Collins A, Fowler A, et al. Orogen styles in the East African Orogen: A review of the Neoproterozoic to Cambrian tectonic evolution. Journal of African Earth Sciences. 2013;**86**:65-106. DOI: 10.1016/j.jafrearsci.2013.06.004

[9] Eliwa HA, Breitkreuz C, El-Gameel K, Khalaf IM. Ediacaran Post-collisional Volcanosedimentary Successions in Egypt. In: Hamimi Z, Arai S, Fowler AR, El-Bialy MZ, editors. The Geology of the Egyptian Nubian Shield: Regional Geology Reviews. Cham: Springer; 2021. DOI: 10.1007/978-3-030-49771-2_7

[10] El-Bialy MZ. Precambrian basement complex of Egypt. Cham: Springer; 2020. pp. 37-79

[11] Akaad MK, El Ramly MF. Geological history and classification of the basement complex in the Central Eastern Desert of Egypt. Geo-Survey Egypt. 1960;**9**:1-24

[12] El Shazly EM. On the Classification of the Precambrian and Other Rocks of Mapnatic Affiliation in Egypt Intern Geol. India; 1964

[13] Ries AC, Shackelton RM, Graham RH, Fitches WR. Pan-African structures, ophiolites and mélange in the Eastern Desert of Egypt, a traverse at 26°

N London. Journal of Geological Society. 1983;**140**:75-95

[14] el Gaby S, el Nady OM, Khudier AA. Tectonic evolution of the basement complex in the Central Eastern Desert of Egypt. Geologtsche Rundschau. 1984;**73**:1019-1036

[15] Ragab AI, El-Alfy Z. Arc–arc collision model and its implications on a proposed classification of the Pan-African rocks of the Eastern Desert of Egypt. MERC Ain Shams University, Earth Science Series. 1996;**10**:89-101

[16] Neumayr P, Hoinkes G, Puhl J, Mogessie A, Khudeir AA. The Meatiq dome (Eastern Desert, Egypt) a Precambrian metamorphic core complex: petrological and geological evidence. Journal of Metamorphic Geology. 1998;**16**:259-279

[17] Habib ME, Ahmed AA, El-Nady OM. Two orogenies in the Meatiq area of the Central Eastern Desert Egypt. Precambrian Research. 1985;**30**:83-111

[18] Bennett JD, Mosley PN. Tiered tectonics and evolution, E Desert and Sinai, Egypt. In: Matheis G, Schandelmeier H, editors. Current Research in African Earth Sciences. Rotterdam: Balkema; 1987. pp. 79-82

[19] El-Gaby S, List FK, Tehrani R. Geology, evolution and metallogenesis of the Pan-African Belt in Egypt. In: El-Gaby S, Greiling RO, editors. The Pan-African belt of Northeast Africa and adjacent areas. Weisbaden: Vieweg & Sohn; 1988. pp. 17-68

[20] Shackleton RM, Ries AC, Graham RH, Fitches WR. Late Precambrian ophiolitic melange in the Eastern Desert of Egypt. Nature. 1980;**285**:472-474

[21] Bregar M, Bauernhofer A, Pelz K, Kloetzli U, Fritz H, Neumayr P. A late Neoproterozoic magmatic core complex in the Eastern Desert of Egypt; emplacement of granitoids in a wrench-tectonic setting. Precambrian Research. 2002;**118**:59-82

[22] Stern RJ. Late Precambrian ensimatic volcanism in the Central Eastern desert of Egypt. San Diego: University of California; 1979. p. 210

[23] Stern RJ. Petrogenesis and tectonic setting of Late Precambrian ensimatic volcanic rocks, Central Eastern Desert of Egypt. Precambrian Research. 1981;**16**:195-230

[24] Sturchio NC, Sultan M, Batiza R. Geology and origin of Meatiq dome, Egypt: A Precambrian metamorphic core complex? Geology. 1983;**11**:72-76

[25] Stern RJ, Hedge CE. Geochronologic and isotopic constraints on late Precambrian crustal evolution in the Eastern Desert of Egypt. American Journal of Science. 1985;**285**:97-127

[26] Johnson P, Andresen A, Collins A, Fowler A-R, Fritz H, Ghebreab W, et al. Late Cryogenian–Ediacaran history of the Arabian–Nubian Shield: A review of depositional, plutonic, structural, and tectonic events in the closing stages of the northern East African Orogen. Journal of African Earth Sciences. 2011;**61**:167-232. DOI: 10.1016/j.jafrearsci.2011.07.003

[27] Thurmond AK, Stern RJ, Abdelsalam MG, Nielsen KC, Abdeen MM, Hinz E. The Nubian Swell. Journal of African Earth Sciences. 2004;**39**(3-5):P401-P407. DOI: 10.1016/j.jafrearsci.2004.07.027

[28] Stern RJ, Ali K. Crustal evolution of the Egyptian Precambrian rocks. In: Hamimi Z, El-Barkooky A,

Martínez Frías J, Fritz H, Abd El-Rahman Y, editors. The Geology of Egypt. Switzerland: Springer Nature; 2020. pp. 131-151

[29] Hume WF. Geology of Egypt, V. II, part III. Egypt: Geological Survey; 1937

[30] Newberry P. Beni Hassan II: Archeological Survey of Egypt. London: Kegan Paul; 1893. p. 8

[31] Meyer C, Heidorn L. Three Seasons at Bir Umm Fawakhir in the Central Eastern Desert. Leiden: Research School CNWS; 1998. pp. 197-209

[32] Roubet C. The grinding stones of site E-78-3 Wadi Kubbaniye. The Prehistory of Wadi Kubbaniye. 1989;**3**:473-489

[33] Harrell JA, Brown VM. The Oldest Surviving Topographical Map from Ancient Egypt: (Turin Papyri 1879, 1899, and 1969). Journal of the American Research Center in Egypt. 1992;**29**:81-105. DOI: 10.2307/40000486

[34] Klemm D, Klemm R, Murr A. Gold of the Pharaohs–6000 years of gold mining in Egypt and Nubia. Journal of African Earth Sciences. 2001;**33**(3):643-659

[35] Klemm R, Klemm D. Gold and Gold Mining in Ancient Egypt and Nubia, Geoarchaeology of the Ancient Gold Mining Sites in the Egyptian and Sudanese Eastern Deserts. Berlin, Heidelberg: Springer-Verlag; 2013. p. 663

[36] Mond R, Winkler HA. Rock drawings of Southern upper Egypt. ii. jstor. 1940

[37] Botros NS. A new classification of the gold deposits of Egypt. Ore Geology Reviews. 2004;**25**(1):1-37

[38] Botros NS. Stratiform gold deposits in a Proterozoic BIF, Abu Marawat area, Eastern Desert, Egypt. In: Eleventh Symposium on Precambrian and Development. Cairo; 1995

[39] Basem AZ, Johnson PR, Goldfarb RJ, Klemm DD. Orogenic gold in the Egyptian Eastern Desert: Widespread gold mineralization in the late stages of Neoproterozoic orogeny. Gondwana Research. 2019;**75**:184-217. DOI: 10.1016/j.gr.2019.06.002

[40] El Alfy EM. Report on the Barramiya gold mine, Eastern Desert. Qusseir Phosphate Co., Cairo.EMRA, Internal reports No. 18 / 68-39 / 77-16 / 78-71 / 2003—Al Alfy, E.M., 1951). 1951

[41] Foda S. Report on Prospecting Work Carried Out by the Mechanical Expedition at Atud, Eastern Desert, 1953-1954. Cairo: Mines and Quarries Dept; 1958

[42] Rogers BM. An inspection of the ancient gold mining region of the South–Eastern Desert between the Red Sea and the Nile. February 23rd to March 18th 1914. Unpublished report. 1914

[43] Moharram O, El-Ramly MF, Amer AF, Ivanov SS, Gachechiladze DZ. Studies on Some Mineral Deposits of Egypt. Egypt: Geol. Surv, of Egypt; 1970

[44] Egyptian mineral resources authority 2014, internal report in, No. 35/2014

[45] Zaghlol K. Structural Control, and Petrology of Gold Occurrences of the Gebel Hamrat Ghannam—Gebel Atawi Stretch, South Quseir, Central Eastern Desert, Egypt. Egypt: Cairo University; 2022

[46] Kochine G, Bassyuni F. Mineral Resources of the UAR, Part I, Metallic Minerals. Egypt: Int. Rep. Geol. Surv; 1968

[47] EI Ramly MF, Ivanov SS, Nochm GC. The Occurence of Gold in the Eastemdesert of Egypt in Studies on Some Mineral Deposits of Egypt. Part I, Sec. A, Mettallic Minerals. Egypt: Geol. Surv; 1970. pp. 53-63

[48] Sabet AH, Bessonenko VV, Bykov BA. The intrusive complexes of the central Eastern Desert of Egypt. Annals of Geological Survey Egypt. 1976;**1976**:12

[49] Sabet A, Bordonosov V. "The gold ore formations in the Eastern Desert of Egypt." ("New Occurrence of Gold Mineralization at the Eastern Part of Wadi El ..."). Annals of Geological Survey Egypt. 1984;**16**:35-42

[50] Botros NS. Metallogeny of gold in relation to the evolution of the Nubian Shield in Egypt. Ore Geology Reviews. 2002;**19**:137-164

[51] El-Aref M, El-Rahman A, Yasser Z, Basem S, Adel H, Hassan A, et al. Mineral Resources in Egypt (I): Metallic Ores. In: Hamimi Z et al., editors. The Geology of Egypt. ("Prof. Dr. rer. Nat. F. Christoph Breitkreuz | TU Bergakademie Freiberg") Regional Geology Reviews. Cham: Springer; 2020. pp. 521-587. DOI: 10.1007/978-3-030-15265-9_14

Chapter 3

Concentrated Hydrochloric Acid Leaching of Greenland Steenstrupine to Obviate Silica Gel Formation

Damien Krebs and Domenic Furfaro

Abstract

The Ilimaussaq complex in Greenland contains a rare earth bearing mineral called Steenstrupine. This mineral is a complex sodium rare earth phospho-silicate which also contains significant uranium and thorium. The mineral can be beneficiated via froth flotation to produce a mineral concentrate ranging between 15 and 23% rare earth oxide. Leaching of the mineral concentrate is required to dissolve the contained values and recover them using hydrometallurgy. Steenstrupine contains high amounts of acid soluble silica which can result in the formation of silica gel during leaching. Laboratory scale testwork was performed to determine which leaching conditions offer the control of silica and high extraction of values such as rare earth elements and uranium. A range of leach parameters where investigated to determine which are most significant to leach performance. Optimised parameters consisting of acid strength, residence time and nature of the operation were determined as significant. In conclusion the operating range identified produces high rare earth extractions while yielding a leach residue with suitable solid liquid separation performance. This process may be applied to other rare earth bearing minerals which contain high proportions of soluble silica.

Keywords: rare earth elements, leaching, silica gel, hydrometallurgy, steenstrupine, hydrochloric acid leaching

1. Introduction

1.1 Project information

The Ilimaussaq complex is located within the Kommune Kujalleq, the municipality of southern Greenland (**Figure 1**). It is a peralkaline intrusion which contains rocks which are extremely enriched in sodium, iron and high fields strength elements including rare earth elements.

Within the complex is the Kvanefjeld deposit which consists of a measured resource (JORC 2012) of 143 million tonnes @ 1.2% REO, 303 ppm U3O8 and 0.24% zinc [1].

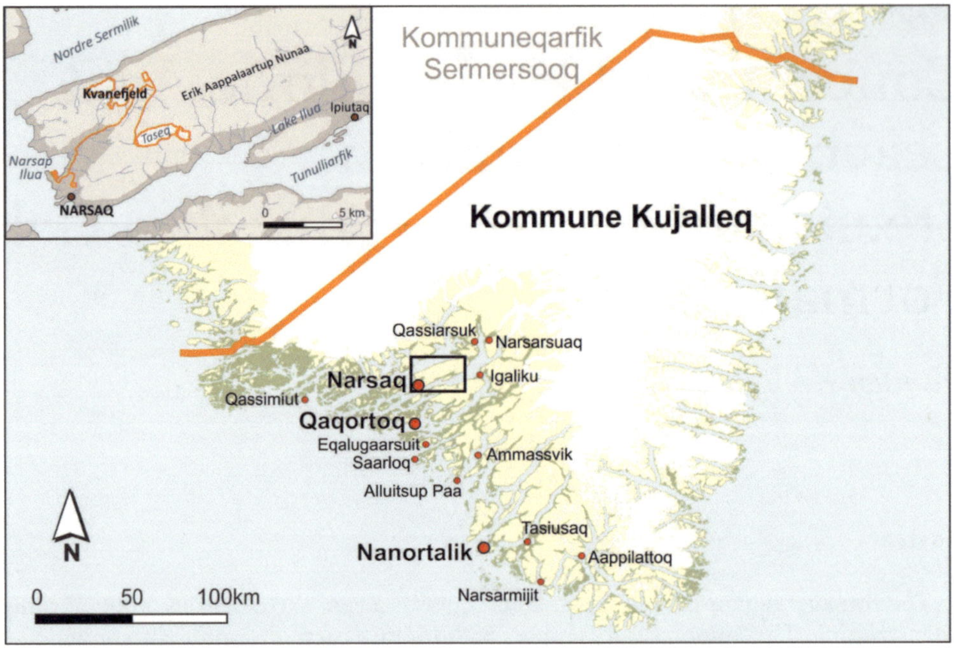

Figure 1.
Map of Kommune Kualleq showing local population centres and project infrastructure.

Greenland Minerals Limited (GML) proposes to develop a mine and integrated minerals processing facility for Kvanefjeld. In addition to producing significant quantities of REE products, the project will also produce, as by-products, small but commercially valuable quantities of uranium, zinc concentrate and fluorspar.

The rare earth elements, in the Kvanefjeld deposit, are contained predominantly in the mineral steenstrupine, which has the chemical formula provided below:

$$Na_{6.7}H_xCa(REE)_6(Th, U)_{0.5}(Mn_{1.6}Fe_{1.8}Zr_{0.3}Ti_{0.1}, Al_{0.2})(P_{4.3}Si_{1.7})O_{24})(F, OH).nH_2O.$$

It contains all 15 rare earth elements, radionuclides, phosphorous and soluble silica. Rare earths are also found to a lesser extent in the minerals townendite, monazite and vitusite [2].

Beneficiation of steenstrupine from the gangue minerals is possible via the froth flotation method. Froth flotation can be used to produce a steenstrupine mineral concentrate which typically grades 14–23% rare earth oxide [3, 4]. Gangue minerals such as felspars, feldspathoids, agerine and amphiboles are extensively rejected and report to the tailings stream. This paper will focus on the hydrometallurgical treatment of flotation concentrates as this is the typical process for rare earth recovery. The inventive aspect is in the definition of process variables which result in a leaching process which produces high extractions while exhibiting good solid liquid separation characteristics of leach residues.

1.2 Mineral leaching issues

As steenstrupine is a very rare mineral there is little research and commercial precedent for it's treatment [5]. It contains a high degree of acid soluble silica and nonrefractory rare earth elements which makes it's leaching characteristics unusual. Historical leaching tests were performed on ore samples which contained steenstrupine [6]. This early work showed steenstrupine and some associated gangue minerals are susceptible to silica gelling. Steenstrupine and it's associated gangue silicate minerals contain acid soluble silica which dissolves into solution with acid leaching [7, 8].

Other researchers have found silica issues with another alkaline rare earth mineral Eudialyte which also contains significant soluble silica [8–10]. Silica soluble minerals which leach under acidic conditions will release silica into solution. If not carefully controlled, dissolved silica can precipitate rapidly forming a network structure in solution resulting in the solution gelling and exhibiting poor solid liquid separation [11–13].

Some leaching methods have been applied to steenstrupine which include both sulphuric and hydrochloric acids [1, 3, 7]. In 2012 GML developed a method which uses a two stage sulphuric acid leach to extract the values and control silica [7]. Since 2017, GML has been developing a different leaching method which utilises concentrated hydrochloric acid in a single stage leaching process [14]. Previous steenstrupine hydrochloric leach tests investigated low solids loading, relatively dilute acid solution and only batch operation [8]. This was effective in leaching rare earths from steenstrupine however a pyrometallurgical pre-treatment was required. In addition the leach density was moderately dilute.

This paper presents the results of investigating the leach performance at higher solids loading and concentrated acid concentrations without a pre-treatment step. The viability of the process is demonstrated by comparing batch leach results with those of continuous leach results.

2. Samples utilised

2.1 Low grade flotation concentrate

The low grade flotation concentrate was generated from Kvanefjeld ore (Lujavrite) during continuous flotation pilot plant testwork. Mineralogical examination (QEM-SEM) identified steenstrupine, arfvedsonite, aegirine, analcime, K feldspar and plagioclase as the major components of the concentrate [3]. Elemental analysis of the concentrate used a combination of fusion inductively coupled plasma optical emission spectroscopy (ICP-OES) and inductively coupled plasma mass spectroscopy (ICP-MS). The chemical analysis was performed by the commercial SGS Laboratory in Perth, Australia who are certified and ISO accredited. The accuracy of the assays is considered very good with an accuracy of +/− 5% (**Table 1**).

Acid soluble silicon concentration in the concentrate was determined by diagnostic leaching of concentrate with 70 g/L Hydrochloric acid (HCl) for 2 hours at ambient

La	Ce	Nd	Pr	Y	U
2.08	3.74	1.30	0.40	0.68	0.18
Al	**Fe**	**Na**	**P**	**Si**	**Th**
3.91	9.27	8.29	2.19	19.5	0.52

Table 1.
Low grade concentrate assays (% of mass).

La	Ce	Nd	Pr	Y	U
4.00	8.00	2.60	0.80	1.40	0.34
Al	**Fe**	**Na**	**P**	**Si**	**Th**
1.5	6.00	8.50	4.00	15.0	1.00

Table 2.
High grade concentrate assays (% of mass).

temperature at 1 and 3% w/w solids. The leach liquor was recovered and filtered through 0.45 microns before diluting 1:10 in deionised water and submitting for assay by Inductively Coupled Plasma – Optical Emission Spectrometry (ICP-OES). The low solids density resulted in low silicon tenors in solution which prevented silica gelling issues. The acid soluble silicon concentration in the concentrate was determined to be 38 g of silicon per kilogramme of concentrate.

2.2 High grade flotation concentrate

Batch testwork on Kvanefjeld ore using an improved flotation process [3] generated a limited quantity of high-grade flotation concentrate which was available for leach testwork. The concentrate elemental composition was determined using a combination of sodium fusion and inductively coupled plasma. The elemental assays are shown in **Table 2**.

The acid soluble silicon concentration in the concentrate was determined to be 32 g of silicon per kilogramme of concentrate solids.

3. Leaching methods

3.1 Batch leach tests

High acidity and high temperature are known to assist with the precipitation of silica in a solid crystalline form rather than polymerising to form silica gel [11–13]. Higher leach slurry densities offer higher acid concentrations in solution and lower mass throughputs. Lower mass throughputs reduce construction costs for the leaching equipment. Concentrated HCl was selected to produce a high solids density and high acidity in the leach. The use of concentrated HCl will increase the temperature of the leach due to the heat generated by acid dilution.

The program focused on leaching low grade concentrate under various atmospheric conditions with hydrochloric acid to determine metal dissolution and the relative filtration performance of the resultant leached slurry. The general acid leach test procedure is shown in **Figure 2**. All acid doses are quoted in grams of anhydrous HCl per gram of dry concentrate.

The concentrate was leached by manually mixing concentrate and concentrated HCl (20–29% w/w HCl) for 15–30 minutes. Dilute HCl (10 g/L HCl) is then added to generate a slurry which is sufficiently fluid to enable it to be poured onto a vacuum filter. The quantity of 10 g/L HCl added to the leach slurry was varied so that the mass of slurry and the slurry % solids to the filter was constant regardless of the acid concentration or acid dose used in the leach stage. The diluted slurry was then vacuum

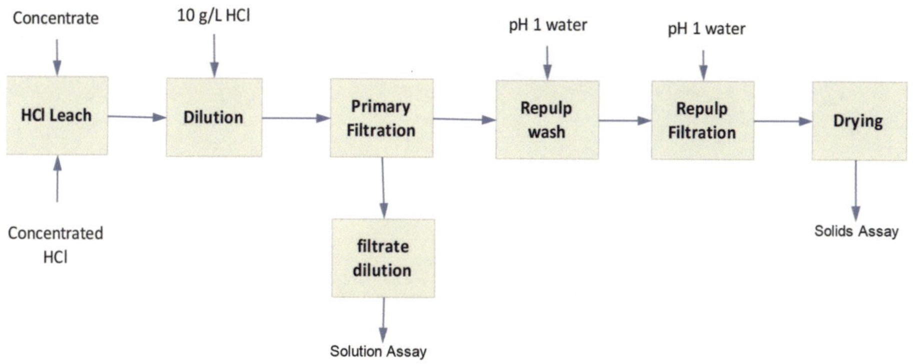

Figure 2.
Batch laboratory work flow.

filtered and the cake form time determined. Filter cake form time was used as a proxy for filtration rate.

The filtrate was collected and a subsample immediately diluted 1:10 with 1 M HCl, to stabilise the sample prior to submitting for elemental analysis. The filter cake was dispersed in D.I water which had been acidified to pH 1 (HCl) and vacuum filtered. Once the surface moisture had disappeared from the filter cake, additional pH 1 wash water was applied to the cake to remove entrained dissolved salts. The solids were then dried to constant mass at 80°C and submitted for elemental analysis.

3.2 Continuous leach tests

Continuous leach tests were conducted once suitable conditions for concentrate leaching were identified from the batch testwork program. Dry concentrate was continuously fed (vibratory feeder) to a 5.8 L stirred baffled reactor. Concentrated (25% w/w) HCl was continuously pumped in via a peristaltic pump to achieve the required acid dose. The reactor temperature was maintained at 80°C using indirect steam through titanium steam coil. A thermocouple was inserted into the slurry and indirect steam added through a solenoid valve connected to an automatic controller.

The reactor had a nominal residence time of between 69 and 97 minutes depending on feed flowrates. Samples were taken from the reactor overflow and prepared for assay as previously described.

3.3 Continuous leach tests

Analysis of the solid samples was performed using a four-acid digest followed by ICP-OES and ICP-MS finishes. The digest solutions were appropriately diluted before being analysed. Solid samples were also analysed by sodium peroxide fusion digest, followed by an analysis of the resulting solutions by ICP-OES and ICP-MS. Solids were assayed using sodium fusion to ensure any refractory minerals were dissolved. Good agreement was achieved between the four acid digest and peroxide fusion methods and suggests that any refractory minerals present in the sample was minimal.

Elemental assays of liquors were determined by ICP-OES and ICP-MS. The liquors were diluted 1:10 with 1 M HCl prior to submission to stabilise silicon in solution.

Free acid was determined by titration with standard NaOH using Calcium EDTA as a complexing agent, the procedure was modified from that described in [15] and

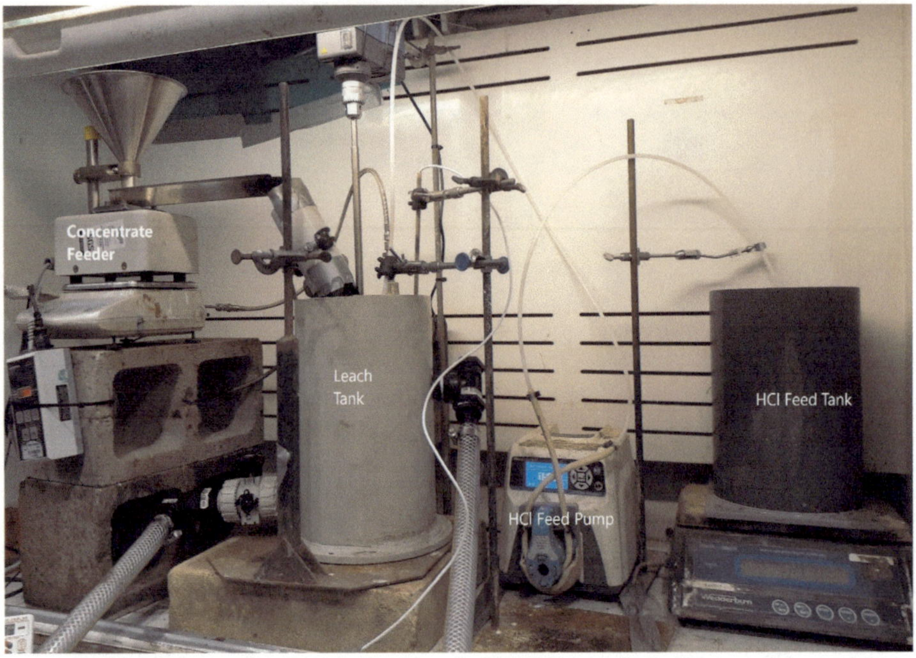

Figure 3.
Continuous leach set up.

required correction for phosphoric acid in the sample, based on the phosphorous assay of the solution (**Figure 3**).

4. Results and discussion

4.1 Batch test results

Four tests were performed to examine the impact of acid dose on metal dissolution and solid liquid separation of the leached slurry. In each test concentrate was mixed with 29% w/w HCl for 30 minutes. The impact of acid dose on filtration rates is provided in **Table 3**.

The metal leach extraction for the four batch tests are shown in **Table 4**. They show very high extraction of rare earth elements. Gangue elements such as aluminium and iron are lower. This is due to the deportment of these gangue metals in the

Test	Acid Dose (g HCl/g con)	Cake Form Time (sec)	Terminal Free Acid (g/L HCl)	Gell Formed
1	0.48	96	119	No
2	0.38	180	89	No
3	0.35	282	68	Yes
4	0.32	642	47	Yes

Table 3.
Acid dose and solid liquid separation performance.

Test	Al	Fe	Na	P	Th	La	Ce	Nd	Y	U
1	64	13	64	94	90	96	98	98	96	84
2	63	11	64	93	87	95	97	97	94	87
3	63	12	64	93	84	95	97	97	95	84
4	63	11	63	91	79	95	96	96	95	85

Table 4.
Low grade concentrate metal dissolution for batch tests.

minerals leached. Most of the iron is found in the black amphibole arfvedsonite which appears to be refractory in the leach process. The refractory aluminium and sodium may be present in sodic feldspar minerals which do not leach extensively under these leach conditions.

Due to the high slurry density it was important to observe the viscosity of the slurry mixture after the addition of concentrate and acid. For the first 1–2 minutes of mixing the mixture would remain fluid, however after this period the free liquid would disappear and the viscosity of the mixture increased to the consistency of damp solids. The damp solid became hot and with continued mixing, formed a viscous paste which progressively became more fluid with time. These observations have also been reported when leaching eudialyte concentrates [16] with strong acid.

Figure 4 shows there is very little impact of acid dose on the dissolution of metals except thorium over the range investigated. Under all acid doses near complete extraction of rare earths and uranium was observed.

The cake form time (filtration rate) shows a strong correlation with acid dose. As the acid dose increased the cake form time decreased, demonstrating an improvement in filtration. At the higher acid doses no gel formation was observed. Both lower acid doses resulted in gel formation indicating that lower acid conditions result in favourable conditions for silica polymerisation. At higher acid doses and therefore higher acid strength in solution may result in a more crystalline silica precipitate.

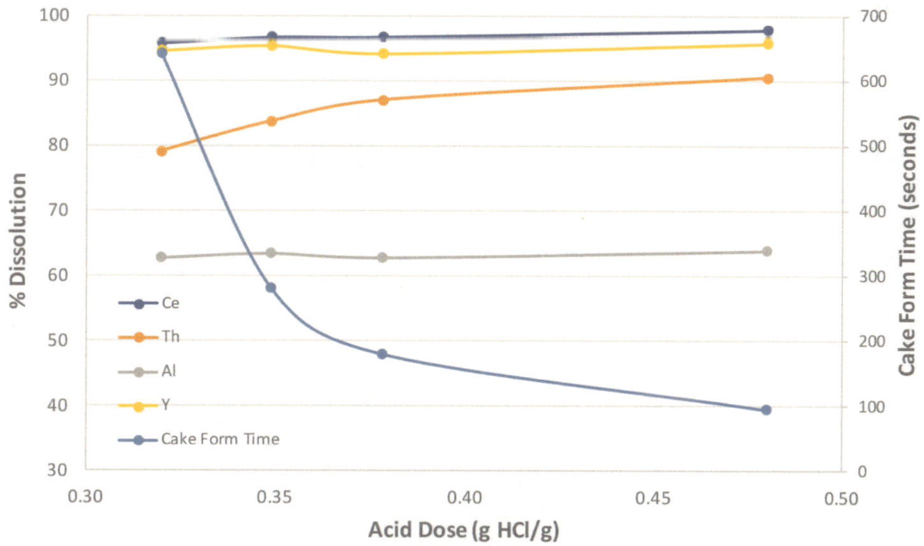

Figure 4.
Acid dose effects on dissolution and filtration.

The reduced filtration rate and the appearance of silica gel at the lower acid doses are all consistent with reduced rates of silica coagulation [12, 13, 17]. The rate of silica coagulation from solution has been shown to be strongly influenced by a number of factors including temperature, the presence of metals in solution and free acid [17, 18]. Generally, as temperature increases the rate of silica precipitation increases while the presence of aluminium in some systems appears to increase the rate of silica precipitation. In these series of tests, the aluminium tenor and the temperature were similar over the acid dose range investigated. As the terminal free acid increased the solid liquid separation improved. The results are consistent with improved silica coagulation at higher acid concentrations resulting in improved filtration.

Over 99% of the acid soluble silicon had precipitated under all the test conditions. Terminal free acid concentrations down to 47 g/L HCl are still capable of achieving high rare earth dissolution. A stoichiometric acid balance was performed to determine the elemental acid consumption. Most of the acid (~60%) was consumed in the dissolution of Al, Na and Fe. A significant proportion (15–25%) of the acid added is not consumed in the reaction and remains in solution as free HCl. The majority of the balance of the hydrochloric acid dose is consumed by rare earth dissolution.

4.2 Impact of addition method

Previous tests were typical batch tests where all the acid and concentrate are mixed together at the start of a test. While this is a convenient procedure it can produce different results to a continuous process which represents a commercial operation. Continuous leach tests where precipitation of reactants occurs can result in significantly more heterogeneous precipitation. Heterogeneous precipitation products can have different particle size and morphology compared to homogeneous precipitation products [19].

Four tests were performed to examine acid addition methods and more continuous leach conditions. Continuous operation was simulated by dosing the reactants to the leach in staged doses during the leach. The acid dose (0.38 g/g), acid concentration (29% w/w HCl), average leach time (30 minutes) and temperature (ambient) were kept constant between tests.

The additions to the reactor were as follows:

Test 9–100%, all acid and concentrate added at the start (baseline test).

Test 11–80% of the concentrate added at the start, remainder after 10 minutes.

Test 14–40% of the concentrate added at the start, remainder after 10 minutes.

Test 15 - a third of the acid and concentrate added every 5 minutes to the reactor.

Table 5 shows that the method of feed addition had little impact on metal extraction with constant acid dose. The filtration of the residues was significantly impacted by the timing of acid addition as observed in **Figure 5**.

The smaller incremental additions to the reactor are more representative of a continuous leach reactor. The results show that cake form time decreases (better filtration) as the proportion of sample added to the reactor at the start of the test fell. These are consistent with increases in heterogeneous precipitation and particle growth expectations. As incremental additions are made to the reactor; silica dissolves and re-precipitates on precipitated silica from earlier reactor additions, increasing silica precipitation particle size. Incremental feed addition to the batch reactor results in a substantial improvement in the filtration properties of the leach slurry.

Test	Al	Fe	Na	P	Th	La	Ce	Nd	Y	U	CFT* (sec)
1	63	12	64	93	84	95	97	97	95	84	245
2	63	12	64	92	84	95	96	97	94	83	220
3	63	13	63	92	85	95	96	96	93	82	106
4	63	12	63	92	85	95	96	96	93	82	52

**CFT = Cake Form time in seconds.*

Table 5.
Impact of acid addition method.

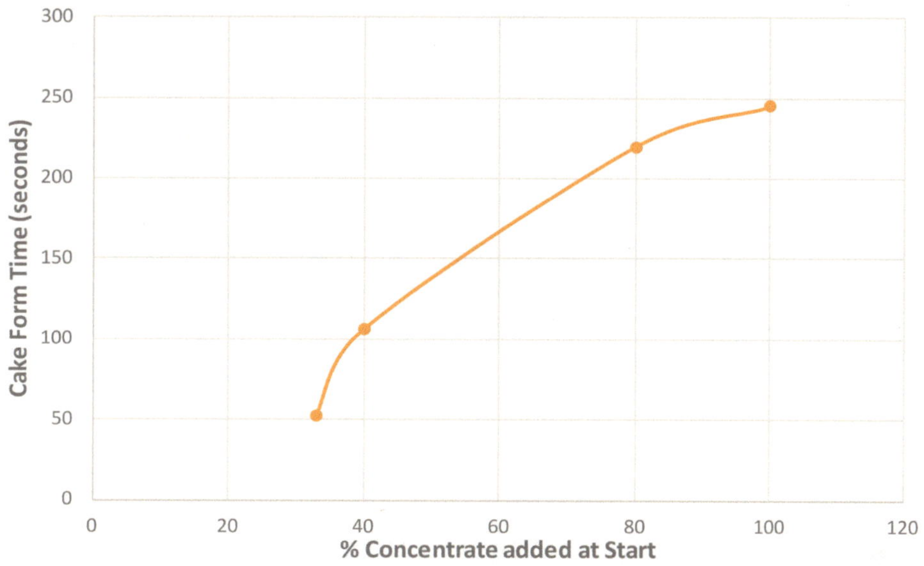

Figure 5.
Dissolution and filtration versus acid dose.

4.3 Impact of acid concentration

Three tests were completed where the acid concentration was varied while maintaining a constant acid dose of 0.38 g HCl /g. The leach mixture was maintained at 55°C in a water bath. To better reproduce the conditions observed in continuous operation the acid and concentrate were added incrementally to a stirred reactor. The first two tests were under the same conditions except the second test used a higher acid concentration (but same acid dose). The third test was a repeat of the first test except a longer residence time (slower addition rate) was employed. The results of these tests are shown in **Tables 6** and 7.

All tests gave similar metal extractions except for thorium which shows a significant drop in dissolution at the lower terminal free acid concentration (lower acid concentration in feed).

The acid strength had the greatest impact on the physical performance of the leach, with gel formation at the lower acid concentration which increased filtration time. Increasing leach residence time, from 32 to 69 minutes (Test 20) overcame the silica

Test	Residence Time (min)	HCl Conc. (%)	Cake Form Time (sec)	Terminal HCl	Gel Formed
18	32	96	119	No	Yes
19	32	180	89	No	No
20	69	282	68	Yes	No

Table 6.
Impact of acid concentration on solid liquid separation.

Test	Al	Fe	Na	P	Th	La	Ce	Nd	Y	U
18	63	11	62	91	78	95	96	96	97	84
19	63	11	63	92	80	96	97	97	96	83
20	63	12	62	91	88	95	95	96	96	85

Table 7.
Impact of acid concentration on metal extraction (% from concentrate).

gelling issues resulting in improved filtration rate. These observations are consistent with the expected impact of dissolved silica precipitating as a polymerised gel.

The cake form time for test 19 is much shorter (40 seconds) than the comparable batch test (test 2, 180 seconds). This is attributed to the elevated temperature of the acid mixing stage in test 19 and increased heterogeneous silica precipitation when the acid and concentrate are added incrementally to the reactor.

4.4 High grade concentrate test results

Leach tests with high grade concentrate were completed to confirm the general relationships between the various process factors examined in this paper. The concentrate and acid were mixed using the simple batch contact and the incremental addition procedure. The acid mixing stage was maintained at 65°C with the acid added as 25% w/w HCl. Tests were conducted to identify the impact of acid dose, acid concentration and addition method on metal dissolution and filtration rate.

Metal dissolution at the lower acid dose was significantly lower regardless of addition method. The lower grade concentrate gave significantly higher metal dissolution at the 0.38 g HCl/g dose, this is attributed to the refractory nature of the gangue minerals in the low-grade concentrate which do not consume significant quantities of acid. Gangue minerals such as feldspar and arfvedsonite are only significantly present in the lower grade concentrate.

At the lower acid dose **Table 8** reveals the incremental addition method resulted in significantly better filtration which is consistent with results for the low grade concentrate.

At the higher acid dose, rare earths, uranium and thorium recovery increased while aluminium, iron and sodium dissolution were unchanged (**Table 9**). Filtration was significantly better at the higher acid dose while the method of acid addition had no discernible impact on filtration rate. The results are consistent with the higher rate

Acid Addition Type	Acid Dose (g HCl/g)	Cake Form Time (sec)	Terminal HCl (g/L)
Batch	0.38	280	40
Batch	0.53	59	83
Incremental	0.37	75	40
Incremental	0.53	57	84

Table 8.
Impact of acid strength on the solid liquid separation of leached high grade concentrate.

Addition type	Acid dose (g/g)	Elemental extraction in %									
		Al	Fe	Na	P	Th	La	Ce	Nd	Y	U
Batch	0.38	62	17	78	65	11	81	76	73	89	85
Batch	0.53	60	15	76	89	75	94	94	94	96	88
Incremental	0.37	61	15	78	67	20	81	77	74	90	83
Incremental	0.53	61	16	78	89	71	94	95	94	98	89

Table 9.
High grade concentrate metal extraction (% dissolved from concentrate solids).

of silica coagulation at the higher terminal free acid (higher acid dose) giving better filtration rates.

Metal recovery remains unchanged between tests except for thorium which shows a slight drop when a lower acid concentration was used and is attributed to the lower terminal free acid.

4.5 Continuous leach results

Dry low grade concentrate and 25% w/w HCl were continuously added to a stirred reactor maintained at 80°C. Flowrates were controlled to maintain the target acid dose and nominal residence time. The test was undertaken to determine metal dissolution and observe the viscosity of the slurry in the leach reactor.

The first continuous test targeted a nominal residence time of 97 minutes at an acid dose of 0.46 g HCl/g concentrate. The second continuous test decreased the nominal residence time to 69 minutes at an acid dose of 0.39 g HCl/g concentrate. The test results of the continuous tests and the comparable batch test are presented in **Table 10**.

Table 11 shows the metal leach extraction was similar between the continuous and batch leach tests. The filtration rates recorded in **Table 10** were consistently better for the continuous tests at similar acid dose. The continuous leach test observed a significantly lower leach viscosity compared to the batch tests. While the slurry from the batch tests had a toothpaste consistency the slurry from the continuous test was more fluid and was able to freely overflow the reactor.

In batch tests the acid/concentrate mixture initially formed a damp solid which is difficult to agitate effectively. The damp solid did not flow but as mixing continued it became more fluid achieving a consistency similar to toothpaste. If high viscosity is

Test type	Residence time (min)	Acid dose (g HCl/g)	Cake form time (sec)
Continuous	97	0.46	52
Continuous	69	0.39	91
Batch	30	0.47	96
Batch	30	0.38	180

Table 10.
Comparison of continuous and batch leach methods impacts on solid liquid separation.

Test type	Acid dose (g/g)	Elemental extraction in %									
		Al	Fe	Na	P	Th	La	Ce	Nd	Y	U
Continuous	0.46	61	12	62	91	89	94	97	98	98	89
Continuous	0.39	60	13	65	92	81	94	96	96	97	86
Batch	0.47	64	13	64	94	90	96	98	98	96	84
Batch	0.38	63	11	64	93	87	95	97	97	94	87

Table 11.
Comparison of metal extractions from batch and continuous leach conditions.

observed in the continuous operation it will result in more expensive and specialised leaching equipment. The low viscosity slurry observed in the continuous tests will allow for the use of standard Continuously Stirred Tank Reactors (CSTRs) which are low cost, low risk and industry standard (**Figures 6** and 7).

Figure 6.
Viscous leach conditions from batch test.

Figure 7.
Fluid leach conditions from a continuous test.

5. Conclusion

Leaching of steenstrupine concentrate was successfully achieved from both low grade (10% REO) and high grade (21% REO) concentrates. Leaching tests show increasing acid concentrations in solution improved the solid liquid separation and yielded high rare earth extractions. Incremental addition of leach feed to the reactor had little impact on metal dissolution but improved leach residue solid liquid separation significantly. The optimum acid dose was in the range of 0.39 to 0.53 t anhydrous HC/t of dry concentrate. Acid doses lower than 0.35 t anhydrous HCl/t dry concentrate result in poor filterability despite excellent leach extraction.

Filtration characteristics also improved when leaching was performed in a continuous reactor as opposed to a batch reactor. Increasing residence time in the leach also improved filtration characteristics at constant acid dose. Under the same acid regime the batch process gave a very viscous slurry while the continuous system gave a free flowing slurry.

High acid strength, longer residence time and continuous operation improve are the most important factors to produce a leach slurry which is suitable for solid liquid separation.

Further research could be performed to examine the impact of temperature on the solid liquid separation efficiency.

Acknowledgements

The authors would like to thank Greenland Minerals for their permission to publish this paper.

Author details

Damien Krebs* and Domenic Furfaro
Greenland Minerals Ltd, Perth, Australia

*Address all correspondence to: damien@ggg.gl

References

[1] Marks AW et al. The Ilimaussaq alkaline complex, South Greenland. In: Charlier B, Namur O, Latypov R, Tegner C, editors. Layered Intrusions. Springer Geology. Springer, Dordrecht. DOI: 10.1007/978-94-017-9652-1_14

[2] Makovicky M et al. Mineralogical, Radiographic and Uranium Leaching Studies on the Uranium Ore from Kvanefjeld, Ilimaussaq Complex. Southern Greenland, Riso National Laboratory (R-416); 1980

[3] Greenland Minerals Limited. ASX announcement. 2019

[4] Yang J. Studies on Mineralogy and Beneficiation of REE Ores Bulletin of the Geological Society of Finland The 32nd Nordic Geological Winter Meeting. Jan 2016

[5] Zhang J, Edwards C. Mineral decomposition of leaching processes for treating rare earth oxide concentrates. Canadian Metallurgical Quarterly. 2013; **52**(3):243-248

[6] Makovicky M et al. Mineralogical, radiographic and uranium leaching studies on the uranium ore from Kvanefjeld, Ilimaussaq complex. South Greenland: Riso National Laboratory; 1980

[7] Krebs DGI, Furfaro D. Continuous Leaching of Kvanefjeld Concentrate, ALTA Conference Proceedings, Perth, Western Australia, Alta Metallurgical Services Publications. 2014

[8] Yun Y, Stopic S, Friedrich B. Valorization of rare earth elements from a steenstrupine concentrate via a combined hydrometallurgical and pyrometallurgical method. Minerals. 2020;**10**:248

[9] Davris P et al. Leaching of rare earth elements from eudialyte concentrate by suppressing silica gel formation. Minerals Engineering. 2017;**108**:115-122

[10] Lebedev VN. Sulfuric acid technology for processing of eudialyte concentrate. Russian Journal of Applied Chemistry. 2003;**76**(10):1559-1563

[11] Sogabe N, Ikenobu S, Nishiyama F, Sakata Y, Kawahara M. Precipitation of silica in zinc refining process. Pb Zn 2010 - Lead-Zinc 2010 Symposium, Held in Conjunction with COM 2010. 2010. pp. 553-563

[12] Terry B. The acid decomposition of silicate minerals part I. Reactivities and modes of dissolution of silicates. Hydrometallurgy;**10**(2):135-150

[13] Terry B. The acid decomposition of silicate minerals part 2. Hydrometallurgical applications. Hydrometallurgy;**10**(2):151-171

[14] Greenland Minerals Ltd. Enhanced rare earth leaching method established to simplify refinery, reduce capital and operating costs. Australian Stock Exchange announcement, January 15th 2018

[15] Zhaowu Z, Zhenyu B, Zhiqi L. Determination of free acid in rare earth solution by a fixed pH method. Analytical Methods. 2010;**2**:82-85

[16] Voßenkaul D, Birich A, Müller N, Stoltz N, Friedrich B. Hydrometallurgical processing of eudialyte bearing concentrates to recover rare earth elements via low-temperature dry digestion to prevent the silica gel formation. Journal of Sustainable Metallurgy. 2017;**3**:79

[17] Gorrepati E, Wongthahan P, Raha S, Fohler H. Silica precipitation in acidic solutions: Mechanism, pH effect, and salt effect. Langmuir. 2010;**26**(13): 10467-10474

[18] Cooper RMG. Silica precipitation from electrolytic zinc solutions. Curtin University of Technology Thesis. 1998. p. 10906

[19] Demopoulos GP. Aqueous precipitation and crystallization for the production of particulate solids with desired properties. Hydrometallurgy. 2009;**96**:199-214

Chapter 4

Luminescent Materials with *Turn-on* and *Ratiometric* Sensory Response Based on Coordination Compounds of Lanthanides

Claudio Pettinari, Andrei Drozdov and Yuriy Belousov

Abstract

Luminescent lanthanide complexes serve as a unique set of tools for creating sensory materials. The most significant types of sensory response in such materials are the turn-on/off response, when the analyte causes an increase or decrease in the emission intensity, respectively, as well as the ratiometric response, which manifests itself as a change in the ratio of luminescence intensities at different wavelengths. In this paper, we consider two of the most technologically advanced types of luminescent sensor materials based on lanthanide compounds—"turn on" and ratiometric sensors. The production of such materials is not only of importance per their possible application but is especially interesting from a fundamental point of view, since their design requires the implementation of non-trivial solutions.

Keywords: lanthanides, coordination compounds, luminescence, sensors, sensory response mechanism, turn-on sensors, ratiometric sensors

1. Introduction

The lanthanide (Ln) family, due to the features of the electron shell, forms a large number of complexes with a set of luminescent properties that is unique among all non-radioactive elements of the Periodic Table. The lanthanide compounds combine effective luminescence in the UV, visible, or near-IR ranges with narrow spectral lines and constant wavelengths. In addition, in the case of lanthanides, it is especially convenient to drive and control luminescence, which opens the way to the creation of new emerging technologies in a chemical sensorics. In the accompanying paper, we have described the principle of luminescence materials with turn-off sensory response, and here we complete our work by proponing both "turn on" and "ratiometric" sensors. This manuscript also concentrates on the role of elements of the Lanthanide Series in this emerging technology.

2. "Turn on" sensors

There are few publications on "turn on" materials with respect to "turn off" sensors, due to several factors: the rarity of processes accompanied by an increase in luminescence requires a rational design of materials, and at the same time, impurities in analytes can cause quenching processes that level the effect of luminescence enhancement. However, examples of materials with such a response have been growing in recent years, and the described mechanisms can serve as an inspiration for new researchers.

Like quenching, luminescence enhancement can be described by the Stern-Volmer equations, but the definition of the constants is given only in a small number of papers.

The most common "turn on" response mechanisms are:

a. The analyte acts as an antenna, its coordination leads to an increase in the luminescence of lanthanide (Direct Analyte Antenna Function, DAAF).

b. An analyte (anionic or neutral) does not directly act as an antenna, but displaces quencher molecules (usually H_2O) from the lanthanide coordination sphere (solvent-quencher displacement, SQD).

c. The analyte, without destroying the molecular structure of the complex, forms or destroys weak bonds (usually hydrogen), reducing the effectiveness of vibrational quenching control (VQC).

d. The analyte leads to partial destruction of the material, removing the quenching fragment not directly bound to the Ln^{3+} ion (quenching fragment removal, QFR).

e. The analyte suppresses luminescence quenching by PET, FRET, or DEE mechanisms (energy transfer quenching control, ETQC).

f. The analyte affects the electronic structure of the sensor by changing the position of the triplet level (triplet level control, TLC), the effectiveness of sensitization (sensitization control, SC), changing the absorption and excitation spectra (excitation wavelength control, EWC).

An analysis of the literature shows that the SQD strategy (**Figure 1**) is the most often implemented. In this case, the organic solvent molecules or other ligands that do not have their own absorption in the same excitation region of the complex displace water molecules from the coordination sphere of lanthanide. Quenching of Ln^{3+}-centered luminescence through interactions with OH, CH, and NH bonds is caused by the dissipation of the energy of the Ln^{3+} excited state into high-energy stretching vibrations of several neighboring molecules [1, 2]. The efficiency of vibrational quenching on these bonds depends on the energy of the excited state Ln^{3+} and the number of vibrational modes of the X-H bond that cover this energy (**Figure 2**).

For europium ions the required number of OH bonds vibration modes (4) is less than for terbium ions (5), which determines a much more efficient quenching of the luminescence of Eu^{3+}. In the transition to O-D bonds, the number of modes for both ions increases (6 and 5, respectively). As a result, the observed lifetime of $Eu^{3+}_{(aq)}$ is

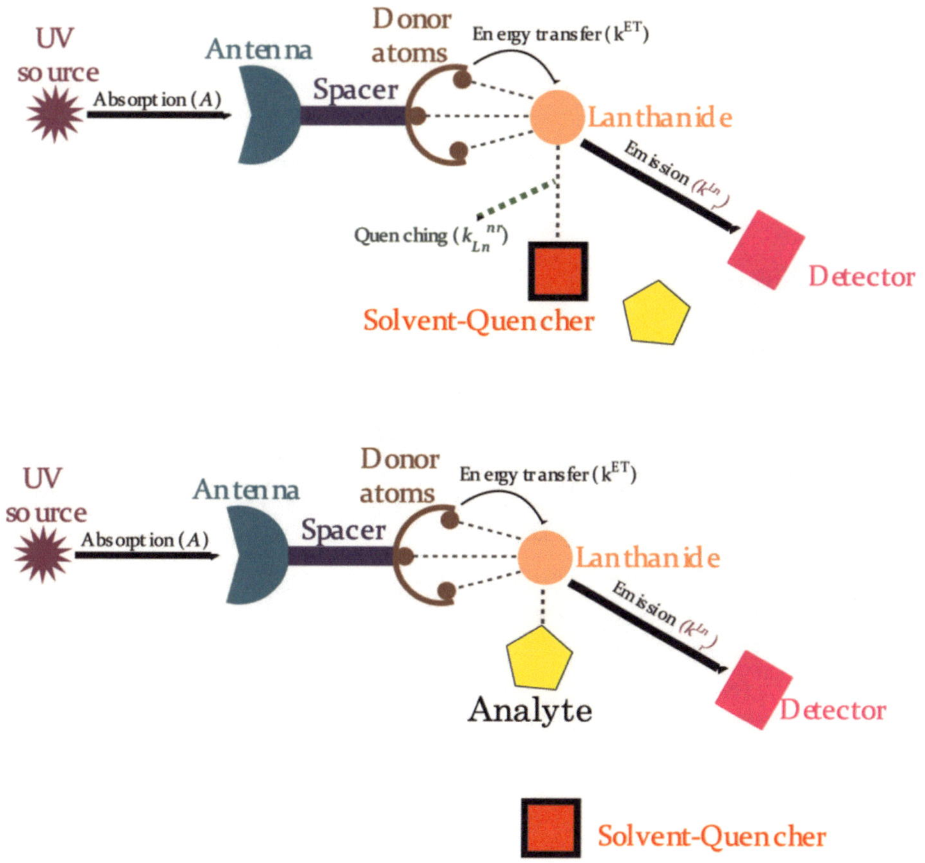

Figure 1.
Substitution of the solvent quencher by the analyte (SQD: solvent-quencher displacement).

shorter than that of $Tb^{3+}_{(aq)}$, and when the medium is replaced by D_2O, the difference is noticeably leveled (see Part 1, **Table 1**). Even greater is the effect of vibrations of these bonds on the emission of IR-emitting ions. The difference in τ_{obs} of various REE ions in protic and deuterated water allows to estimate the number of water molecules in the near coordination sphere of lanthanide according to empirical formulas $q_{Ln} = k \times \left(\frac{1}{\tau_{H_2O}} - \frac{1}{\tau_{D_2O}} - b\right)$, where q_{Ln} is the number of water molecules in the coordination sphere of lanthanide, τ_{H2O} and τ_{D2O} are the observed lifetimes of excited REE states in water and D_2O. k and b parameters are given in **Table 1**. This makes it possible to evaluate the difference in sensitivity for materials based on ions with lower slope values (Nd^{3+}, Eu^{3+}, and Yb^{3+}) and larger ones (Tb^{3+}, Sm^{3+}, and Dy^{3+}). Most of the papers are devoted specifically to europium derivatives, which is not accidental. Neodymium compounds have been studied as sensor materials with less SQD response, but appear to be very promising [7].

An interesting example of this type of sensor is presented in [8], which shows the selectivity of the response with respect to methanol against the background of ethanol and propanol-1. The effect is due to the lock-and-key matching of the channel diameters in the MOF sensor structure with the sizes of the indicated alcohol molecules.

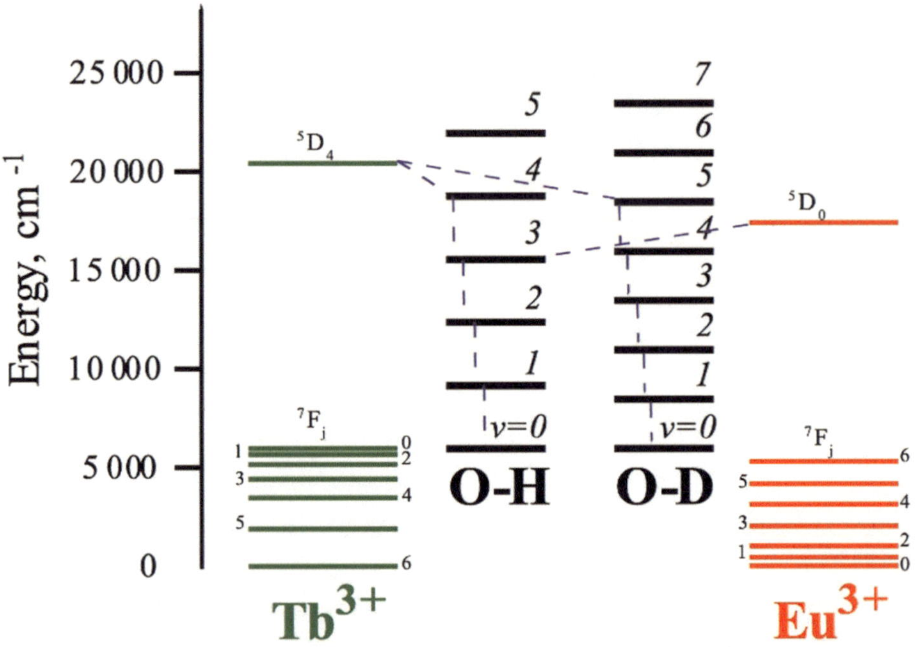

Figure 2.
Nonradiative deactivation of excited states of terbium and europium by multiphonon relaxation on O-H and O-D bonds.

Ln^{3+}	Slope (k)	Intercept (b)	Ref.
Eu	1.11	0.31	[3, 4]
Eu	1.05	0.44	[5]
Tb	5.0	0.06	[3]
Tb	4.03	0.87	[5]
Sm	2.54	0.37	[5]
Dy	2.11	0.60	[5]
Nd	0.358	1.97	[6]
Yb	1.0	0.20	[3]

Table 1.
Slope (k) and intercept (b) in the Ln^{3+} hydration number formulae.

The QFR mechanism (**Figure 3**) has been often implemented using not-luminescent copper-lanthanide heterometallic complexes capable of "donating" copper ions under the action of various analytes, especially those containing sulfur [9–12] and nitrogen [12–14] donor atoms with a high affinity for copper ions. This approach makes it possible to achieve greater selectivity with respect to background ions not strongly interacting with Cu^{2+} ions. In other cases, an analyte with strong oxidizing (ClO^-, [15]) or reductive (ascorbic acid, [16]) power removes the quencher fragment. Other analytes having the same redox properties are expected to show a similar effect.

DOI: http://dx.doi.org/10.5772/intechopen.109189

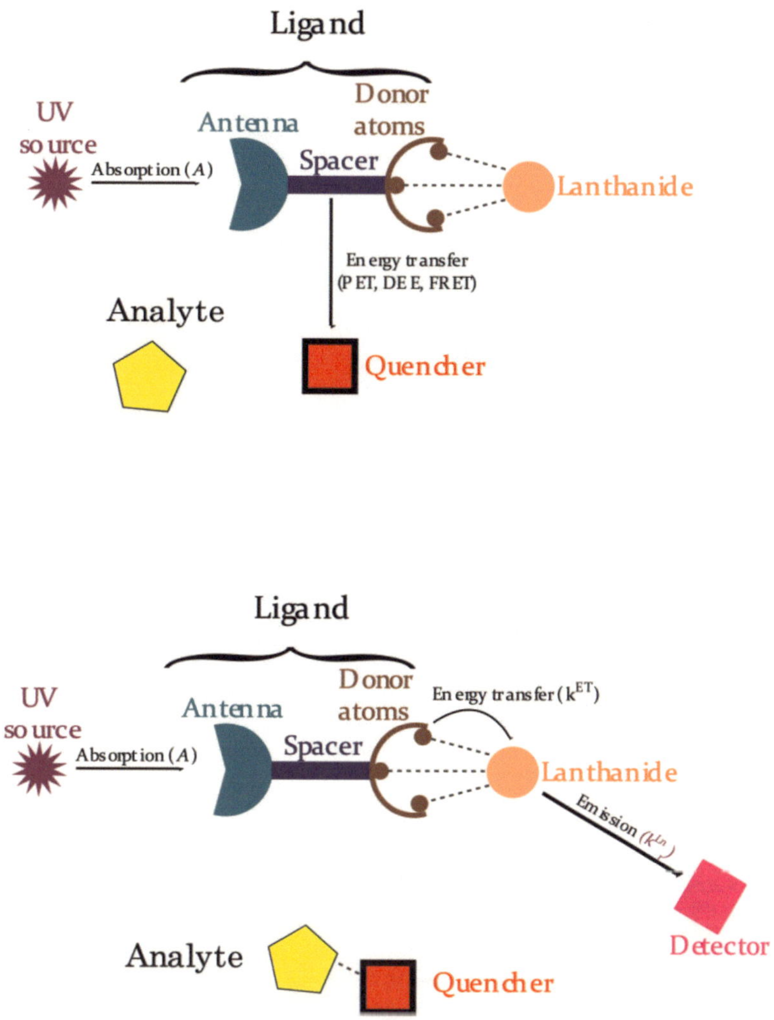

Figure 3.
The quencher fragment removing mechanism (QFR).

The interaction of some analytes with the sensor material leads to a change in the electronic structure, which increases the efficiency of sensitization of lanthanide ions. The detailed mechanism of such enhancement is difficult to determine, but a suppression of the non-radiative relaxation of the singlet and triplet states of the ligands [17], an increase in the energy transfer constant from the ligand to the metal, and a change in the position of the triplet level [18], which affects the efficiency of the reverse transfer, can also make a contribution. These sensor materials showed a response with respect to s- and p-metal cations, as they generally contain a crown-containing fragment [17] or suitable coordination sites determining the material selectivity. The response to gases has been studied much less frequently than the response to analytes in solution; a paper [18] describing the turn-on sensor for NO_2 is of particular interest, two luminescent Eu and Tb complexes being investigated: the Tb complex shows that a reversible sorption of NO_2 leads to a "turn on" response, while for europium, a "turn off" response is observed. The interaction between the

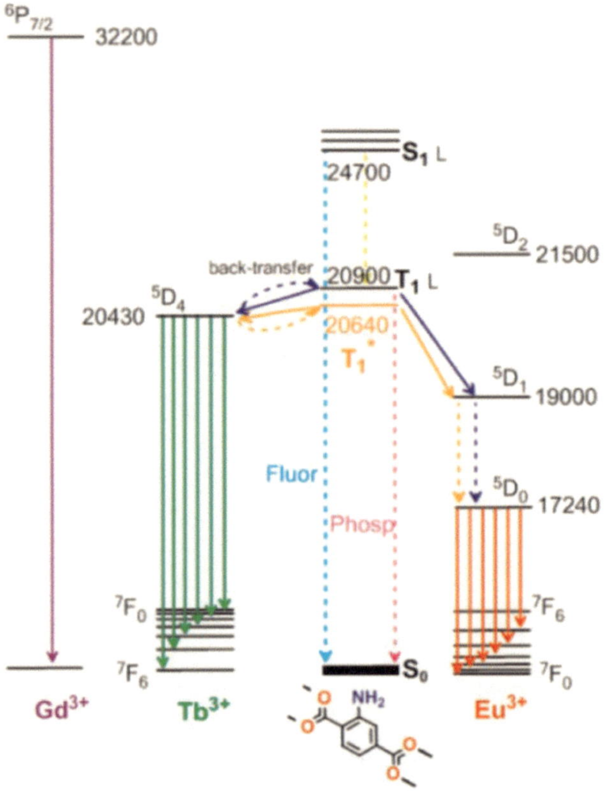

Figure 4.
Jabłoński diagram featuring the ground singlet (S^0), first excited singlet (S^1), and lowest triplet (T^1) states of the ligand together with the relevant atomic levels of Gd^{3+},Tb^{3+}, and Eu^{3+}. Values for the energy levels are given in cm^{-1}. Represented with a permission from Ref. [18].

sensor with the analyte leads to an increase in the energy of the triplet level by ~260 cm^{-1}, which is favorable for energy transfer to a higher resonance level of terbium, but reduces the efficiency for the low-lying level of europium (**Figure 4**).

A sensor material with a positive luminescent response to Cu^{2+} ions has been recently described [19]. Upon addition of up to 4 equivalents of 3d-metal salts (especially Cu^{2+}) an intense band appears in the excitation spectrum associated with intraligand energy transfer. The appearance of this band makes the luminescence excitation more efficient when using the corresponding wavelength, which leads to a more than twofold increase in the quantum yield of Eu^{3+}. This work is a unique example of a turn-on sensor for a d-metal cation.

Finally, the direct antenna function of the analyte (**Figure 5**) can take place if the analyte contains a suitable conjugate system [20, 21]. A not-high selectivity has been showed by such sensors, but the use of a well-defined excitation wavelength, coinciding with the absorption of the analyte, can increase it (**Table 2**).

To determine the correct mechanism of the observed “turn on” response, we propose the following algorithm, which is also relevant for the rational design of such materials:

i. A chemical analysis of the “analyte + sensory material” system to answer the question “whether a new chemical compound is formed” during their interaction.

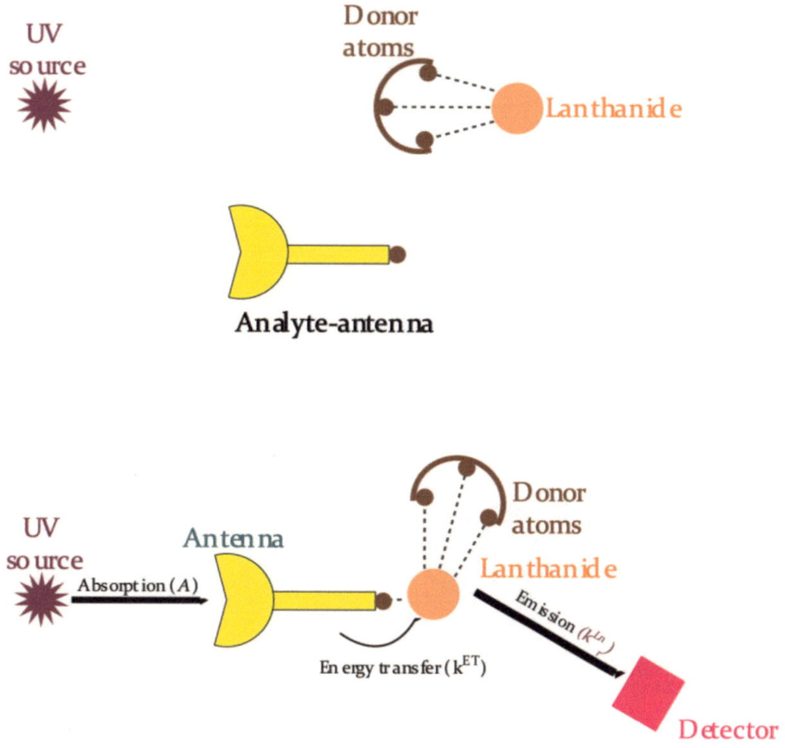

Figure 5.
Direct antenna analyte function mechanism (DAAF).

ii. A study in the case of a positive reply on the elimination of molecules or quenching ions (water, transition metals cations, etc.) occurs. In the case of a positive response, a mechanism such as SQD, QFR, and DAAF can be assumed. The usually observed increase in the τ_{obs} of lanthanide ion is evident in favor of such an assumption. In the case of vibrational quenching, one should keep in mind the significant temperature and isotope sensitivity of the phonon relaxation efficiency, so measurements of τ_{obs} in a deuterated solvent and/or upon cooling are the crucial experiment to confirm the SQD mechanism.

iii. An IR study, in the case of a preserved molecular structure of the sensor to verify the breaking or formation of weak bonds (hydrogen, etc.), which could support a VQC mechanism.

iv. A study confirming that a new ligand (usually an analyte) is coordinated to the sensor and is capable to perform an antenna function. The DAAF mechanism is, in fact, confirmed by the combination of analyte absorption and sensor excitation spectra in the presence and absence of the analyte, as well as measurements of τ_{obs} of Ln^{3+}.

v. A measure of the energy of the triplet level from the phosphorescence spectra of the Gd^{3+} complex in the presence of a chemical interaction between the sensor and the analyte. The change in the energy of the triplet level indicates

Material	Analyte	Media	Linearity range	LOD	Response time	Mechanism	Ref.
Cu-Eu or Cu-Tb heterometallic MOF	S^{2-}	H_2O; HeLa cells		0.01 μM	8–12 min	QFR (Cu^{2+})	[9]
Cu-Tb heterometallic MOF	S^{2-}	H_2O		0.13 μM		QFR (Cu^{2+})	[10]
Hybrid material Eu3+ complex-covered laponite	Glutathione	H_2O	0.5–30 μM	0.162 μM	10 min	QFR (Cu^{2+})	[11]
Hybrid material: Tb^{3+} covered Cu^{2+}-BTC MOF	amyloid β-peptide	H_2O	1–550 nM 5–490 nM	0.3 nM		QFR (Cu^{2+})	[13]
Composite material: carbon quantum dots covered with a H_2DPA. Eu^{3+} and Cu^{2+}	H_2DPA	H_2O	0–10 μM	0.137 μM		QFR (Cu^{2+})	[12]
Tb@duoble hydroxid Ni-Fe	H_2DPA	H_2O	0–12 μM	0.36 μM		SQD+DAAF	[22]
Eu@ duoble hydroxid Ni-Fe	H_2DPA	H_2O	0–12 μM	1.03 μM		SQD+DAAF	
(Cu^{2+}@Tb-MOFs)	Uric acid	H_2O	0–104 μM	0.650μM		QFR (Cu^{2+})	[14]
Tb-MOF	DMSO	H_2O	1.0–100%	1. 1.68%	90s	SQD	[23]
Eu-MOF	MeOH	CH_2Cl_2		<0.03 vol.%		SQD	[8]
Eu-complex nanoparticles	Cipro-floxacin	H_2O	1–40 μM	780 nM	15 min	SQD	[24]
Eu-MOF	Diethyl phosphorochloridate	CH_3CN	15–90 μM		<2 min	SQD	[25]
Eu-MOF	AMF	H_2O		2 μM		SQD	[26]
nanoparticles [Tb(Cit)(H_2O)]	Guanosine-5-mono-phosphate	H_2O	0.15–20 μM	0.1 μM		SQD	[27]
$Tb(HCOO)_3$	l-kynurenine	H_2O + 5% DMF	1–10 μM	1 nM		SQD	[28]
Eu-MOF	$NH_3CH_3NH_2$	Air	(1) 5–350 ppm (2) 0.7–15.5ppm		500–600 s	SQD	[29]
Eu-MOF	NH_3	Air		>28ppm	100–250 s	SQD	[30]
Nd-MOF	F-	CH_3CN				VQC	[7]
Eu(L)-(UIO-67)	ClO^-	H_2O	0.1–5 μM	16nM	5 s	QFR	[15]

DOI: http://dx.doi.org/10.5772/intechopen.109189

Material	Analyte	Media	Linearity range	LOD	Response time	Mechanism	Ref.
Composite material: Eu^{3+} complex in a PMMA membrane reacted with a $KMnO_4$	Ascorbic acid	H_2O	0–35 μM	48 nM	1 min	QFR	[16]
Eu-MOF	Zn^{2+}	MeOH	2–8 mM			VQC	[31]
Eu-MOF	Cd^{2+}	H_2O			2 h	VQC	[32]
EuL TbL EuL' TbL' ML = crown-containing ligands	K^+	MeOH	0–10eq 0–10eq 0–2eq 0–1.7eq			ETQC	[17]
Tb-Zn heterometallic MOF	Mg^{2-}	DMF	10^{-5}–$3{\cdot}10^{-3}$ mM	$1.38{\cdot}10^{-5}$ M $K_{SV} = 1.81{\cdot}10^3$		n./a. probably SC	[33]
Eu-MOF	Cd^{2+}	H_2O	0.2–2.2 μM	$1.1761{\times}10^7$		n./a. probably SC	[34]
Hybrid material: Eu^{3+} covered Zr-MOF Uio-66(Zr)–(BTEC-COOH)$_2$	Cd^{2+}	H_2O	0¯500 μM	0.06 μM	>2 min	SC	[35]
Tb-MOF	Pb^{2+}	EtOH	10^{-8}–10^{-3} M			SC	[36]
Eu-MOF	Pb^{2+}	H_2O	0–0.1 mM K_{SV}=2970 M^{-1}	8.22 μM		n/a	[37]
Tb-MOF	NO_2		1–5 ppm	2 ppm	1 min	TLC	[18]
Eu-MOF	Cu^{2+}	CH_3CN				EWC	[19]
[Tb-MOF	Tryptophan	H_2O pH=4	$2.5\ 10^{-5}$–$2.5\ 10^{-4}$	1 μM	2 h	DAAF	[20]
Nanoparticles of Na[$Gd_{0.88}Tb_{0.12}F_4$]	Dopamine	H_2O	0–10 μM	47 nM	5 min	DAAF	[21]
[Eu-MOF	Al^{3+}	DMF				n/a	[38]
Tb^{3+}-functionalized COF	Ochratoxin A	H_2O	0–10 μM	13.5 nM	10 s	DAAF+SQD	[39]
Tb^{3+} MOF	Acetyl-acetone	H_2O	0–1750 ppm	0.129 ppm		DAAF	[40]
	Aspartic acid	H_2O	0–25 μM	0.025 ppm		DAAF	[40]

Table 2.
Turn-on sensors.

the implementation of a relatively rare TLC mechanism. Similarly, the appearance of a new intense band in the excitation spectrum makes it possible to implement the EWC mechanism.

vi. A combined analysis of both absorption and excitation spectra of the sensor material, which can be used to prove the ETQC mechanism. For a more accurate determination of the latter, a calculation of the energy of the orbitals is required.

3. "Ratiometric" sensors

The third important sensor type is the "ratiometric" one. In materials showing these properties, the signal is estimated as the ratio between luminescence values at two different wavelengths, either affecting the effect of Ln^{3+} ions or affecting both the Ln^{3+} and the organic ligand.

$$Sensory\ Resonse\ (S) = \frac{I_1}{I_2} = \frac{\int_{\lambda_{1(min)}}^{\lambda_{1(max)}} Id\lambda}{\int_{\lambda_{2(min)}}^{\lambda_{2(max)}} Id\lambda}, \quad (1)$$

where $\lambda_{1,2(min/max)}$ -are the initial and final coordinates of the bands in the spectrum involved in the integration.

"Ratiometric" sensors do not have the disadvantage to determine the presence of internal standards as in the case of "turn off" and "turn on" sensors. An unusual uniqueness of ratiometric sensory visibility materials is the naked eye color change of luminescent properties when they are in presence of an analyte.

The possible response mechanisms are similar to the "turn off" and "turn on" signals considered above. In addition, if two different Ln^{3+} ions are simultaneously present in a compound (**Ln-Ln' sensors**), energy transfer between them is possible, the efficiency of which depends on the Ln-Ln distances and other factors [41]; the analyte can affect the transfer efficiency constant. This type of response is defined as Metal-to-Metal Energy Transfer (MMET).

Ratiometric materials generally use the Eu^{3+}/Tb^{3+} bimetallic pair, since these ions have the most efficient luminescence. Other lanthanide-based systems were used only in a small number of cases, for example, Ce^{3+}/Tb^{3+} [42], Dy^{3+}/Eu^{3+} [43], and Eu^{3+}/Yb^{3+} [44] systems. To achieve greater accuracy, as a rule, it is better to employ the most intense lines in the emission spectra, for example, those corresponding to the europium $^5D_0$-$^7F_2$ (~612 nm) and terbium $^5D_4$-$^7F_5$ (~544 nm) transitions, also if this is not entirely correct, because the europium transition $^5D_0$-$^7F_2$ is partially superimposed on the low-intensity transition of terbium $^5D_4$-$^7F_5$ (~620 nm). The sensor response can be calculated using the europium $^5D_0$-$7F_4$ (~700 nm) transition, which in many cases is also very intense and does not overlap with any terbium line, but lies in the region of reduced sensitivity of common photomultipliers [45]. The usual molar fractions ratio of lanthanides with a predominance of terbium ions is caused by the transfer of energy between lanthanide ions.

O_2 in the triplet state can act as a luminescence quencher [46, 47]. The emission of Tb^{3+} ions is quenched by O_2 molecules more efficiently than by Eu^{3+} due to the smaller difference in the energies of ^{3}**T** level of O_2 and the excited state of corresponding Ln^{3+} ions. This principle is the basis for a sensor for gaseous oxygen, which is a mechanical mixture of terbium and europium complexes immobilized on a quartz surface [48].

The possibility to determine, using "ratiometric sensors", small H_2O impurities against the background of organic solvents (for which Karl Fischer titration is usually used), which requires the use of toxic reagents and "capricious" equipment, is of great interest. In addition, "ratiometric" sensors can detect the mixture of light water in and D_2O, which is impossible with Fischer titration and requires expensive mass spectrometers or precision IR spectrometers [49, 50]. Such materials are based on MOF structures containing intra-sphere water molecules. The sensor material is activated by heat treatment in vacuum, after which the obtained anhydrous powder is dispersed in the medium of the investigated solvent. This approach is not accidental: polymeric MOFs are insoluble in most organic solvents, which makes them easy to regenerate and reuse. OH quenching is the basis for the detection of methanol in ethanol and in the form of vapors in air [51], and a similar principle can be extended to CH oscillations in an elegant DMSO impurity sensor in deuterated DMSO-d_6 [44]. Substitution of an OH group [45] or a water molecule [52] in the coordination sphere of lanthanide with F^- ions also suppresses Eu^{3+} quenching more effectively than Tb^{3+}, which was used in the development of fluoride-sensitive sensors.

Mixed-metal MOFs with a statistical distribution of Ln^{3+} ions are commonly used as material for "ratiometric" sensors. However, Tscelykh et al proposed in [53] to use solutions of pentafluorobenzoates or even Ln^{3+} chlorides, since the sensitivity of the sensor is directly dependent on the number of water molecules in the environment of Eu^{3+}. The disadvantages of this approach include the complexity of material regeneration and contamination of the analyzed media, which makes flow analysis impossible and increases its cost.

Moisture sensors are closely related to pH sensor materials. "Ratiometric" pH sensors based on carboxylate MOFs have been also considered [54–56]. In the first two cases, as the pH increases, the color of the luminescence changes from green to red; the relative intensity of Eu^{3+} emission increases. This is explained by the strong pH dependence of the excitation transfer rate constant from Tb^{3+} to Eu^{3+}, which was confirmed by measurements of τ_{obs} in monometallic and bimetallic complexes [55] (in the analysis of kinetic data, certain caution is required due to the non-monoexponential nature of the Eu^{3+} decay curve in the presence of luminescence sensitization by Ln ions [41, 50]). It has been described [56] that as the pH increases, the luminescence color, on the contrary, changes from red to green, that is, the relative intensity of Tb^{3+} luminescence increases. It was shown by the DFT method that the energy of the ^{3}**T*** triplet level increases from ~24.400 to ~26.400 cm^{-1} upon transition from the protonated to the deprotonated form of the ligand. The resonant level of europium is much lower than both of these values, so the sensitization of europium luminescence (17.200 cm^{-1}) does not change its efficiency (τ_{obs}(Eu) also changes slightly with pH). The resonant level of terbium (20.500 cm^{-1}) lies higher, and an increase in the triplet level energy weakens the back transfer of excitation from Tb^{3+} to the ligand. This leads to a significant increase in τ_{obs}(Tb) and in the intensity of Tb^{3+} emission.

The MMET mechanism can be further confirmed by measuring the response in a bimetallic complex and in a mechanical mixture of complexes of two REEs. The decrease or disappearance of the response in the second case indicates the partial or complete participation of the MMET mechanism. A similar approach was used [57] where the MMET mechanism is related to the response to hydrosulfide anions.

Interestingly, the other two analyzed analytes (THF and Ag^+ cations) exhibit different response mechanisms.

The highly selective sensor for a potassium ions, as in the case of the "turn-on" material described above [13], contains diaza-18-crown-6 in the structure, which makes it selective to the tested s-metal cations [58]. It has been shown that the capture of the potassium ion by the crown fragment leads to an increase in the triplet level energy from ~22.400 cm^{-1} to ~23.400 cm^{-1}. This effect enhances the emission of Eu^{3+} to a greater extent than that of Tb^{3+}. In addition, it was found that the efficiency of energy transfer from Tb^{3+} to a Eu^{3+} increases with increasing K^+ concentration, which leads to an increase in the contribution of europium bands in the spectrum.

For Ln-Ln' "ratiometric" sensors, in contrast to the "turn off" and "turn on" materials considered above, the response to conjugated organic analytes has been poorly studied. The sensory response of Eu-Tb bimetallic BTC-MOF film to a number of drugs [59] has been also investigated. The nature of the emission of this material changes significantly in the presence of a number of compounds, in particular, in the presence of coumarin and caffeine. The nature of the response to caffeine is not discussed, and for coumarin, an increase in $k^{Ln}{}_{ET}$ is assumed, which is confirmed by kinetic measurements, but the corresponding quantum chemical studies are not provided.

In the presence of only one lanthanide luminescent center, the "ratiometric" response can still be realized if the ligand is fluorescent or phosphorescent (Ln-L sensors). This is possible if the antenna sensitization efficiency is low, which can be caused by too large or too small energy gap between $^3\mathbf{T}^*$ and the REE resonant level, as well as by a large metal-ligand distance. Possible mechanisms for the occurrence of this type of sensory response include binding of the lanthanide ion caused by decomposition of the parent complex, which weakens the efficiency of antenna sensitization (lanthanide ion ejection, LIE), and direct antenna function of the analyte.

The detection of Hg^{2+} ions was possible by using a composite material containing a luminescent terbium coordination polymer impregnated with a coumarin solution [60]. The response is selective to Hg^{2+} ions with respect to a wide range of s-, p-, and d-metal ions and is associated with the displacement of Tb^{3+} ions from the adenosine monophosphate environment, leading to a decrease in the Tb^{3+} luminescence intensity and to a weakening of its sensitization by coumarin. The driving force behind the displacement reaction of the lanthanide ion into a complex with a lower luminescence intensity can be not only the strength of the "cationic analyte-ligand" bond, but also the formation of a stable "analyte-lanthanide cation" complex. A similar strategy was implemented for phosphate ions [61] and for alkaline phosphatase [62]. In some cases, the response occurs when the ligand is destroyed, for example, in sensors for the determination of HClO [63] or formaldehyde [64]. The vulnerability of these two strategies is the impossibility of a sensor regeneration.

Radiative relaxation by the PET or FRET process can also lead to a response if the analyte has the appropriate LUMO energy. This approach was implemented in a sensor material for the detection of nitrofuranose and furazolidone against the background of other antibiotics in a dysprosium-containing sensor [65]. Trinitrophenol (TNP) similarly blocks the sensitization of terbium luminescence in specifically designed Tb-MOF via PET and FRET mechanisms [66].

As found for Ln-L sensors, the response manifests itself as a drop in the relative efficiency of lanthanide luminescence. The opposite is also possible if the ligand displaces quencher molecules (SQD) and exhibits antenna properties itself (DAAF). An example of this approach is in a study [67–69] describing the response to dipicolinic acid, an important biological marker associated with antrax disease.

Material	Analyte	Media	Response wavelengths λ1, λ2 (nm)	Linearity range	LOD	Mechanism	Ref.
Ln-Ln' materials							
$[Eu_{0.167}Tb_{0.833}(L)(H_2O)_3]$ (1,4-dioxane) MOF	H_2O	D_2O dioxane DMF	543Tb-616Eu	10–120000 ppm		SQD	[49]
$\{[Eu_{0.1}Tb_{0.9}(L)(H_2O)_3]$ MOF	H_2O	D_2O	615Eu-545Tb	0.5–100%		SQD	[50]
		dioxane		0–100%.		SQD	
		CH_3CN		0–60%.		SQD	
$Tb_{0.9711}Eu_{0.0289}(L)_{0.5}$ MOF	H_2O	CH_3CN	543Tb-615Eu	0–2.5 v%		SQD	[73]
$\{[(Eu_{0.02}Dy_{0.98})_2(L)_2]\cdot 2NH_2(CH_3)_2\cdot H_2O\}$ MOF	H_2O	EtOH	416 Dy-614Eu	0–10%	0.1%	SQD	[43]
$[Tb_{0.3}Eu_{0.7}](L)$ MOF	H_2O (vapors)	Air	619Eu-546Tb	20–100%	4.3%	SQD	[74]
$10Eu(C_6F_5COO)_3$+$1Tb(C_6F_5COO)_3$ solution	H_2O	D_2O	612 Eu, 545 Tb	0–50%		SQD	[53]
$Eu_{0.022}Tb_{0.978}$ MOF	H_2O	EtOH	544Tb-614Eu	0–0.67%	0.016%	SQD	[75]
$Eu_{0.2}Tb_{0.8}$MOF	H_2O	EtOH	546Tb-616Eu	10–100%		SQD	[76]
$Eu_{0.2}Tb_{0.8}$MOF	H_2O	EtOH	546Tb-616Eu	0–10%	0.01%	SQD	
$[Eu_{0.05}Tb_{0.95}(L)(H_2O)Cl]$ MOF	pH, H_2O	DMF	547Tb-616Eu	pH 3–11 0–0.8 v% H_2O		SQD	[54]
$Eu_{0.034}Tb_{0.966}(L)_2(C_2O_4)(H_2O)_4$ MOF	pH	H_2O	618Eu-545Tb	pH 3–7		SQD	[55]
$Eu_{0.205}Tb_{0.795}(L)(H_2O)_4\cdot 0.5H_2O$ MOF	pH	H_2O	616Eu-543Tb	pH 4.0–7.5		SQD	[56]
$Eu_{0.4}Tb_{0.6}$ Ciprofloxacin complex + dextran aldehyde (DEX) + chitosan hydrogel	pH	H_2O	615Eu-545Tb	pH 5.5–8.0		SQD	[77]
$\{(Me_2NH_2)[Tb_{0.9}Eu_{0.1}(L)_2]\cdot(L^{76})\cdot(H_2O)_{1.5}\}n$ MOF	MeOH	EtOH, air	545Tb-618Eu	0–1 v%		SQD	[51]

Material	Analyte	Media	Response wavelengths λ1, λ2 (nm)	Linearity range	LOD	Mechanism	Ref.
$Eu_{0.1}Yb_{0.9}(dbm)_{3x}(BPhen)$ solution	DMSO	DMSO-d_6	612Eu-975Yb	0–50 v%		SQD	[44]
$[Eu_{0.1}Tb_{1.9}(FDC)_3(DMF)_2]\cdot 2DMF$	CH_3NHCO	DMF	544Tb-614Eu	0–100 v%			[78]
$[Eu_{0.5}Tb_{1.5}(FDC)_3]$	ethylene glycol (EG)	EG-dioxane mixture	546Tb-616Eu	0–100 v%		SQD	[79]
Hybrid material: mixture of solid Ln(L) complexes (Ln=Tb and Eu) immobilized on glass surface	O_2	Air/N_2	546Tb-616Eu	0–0.2 atm		ET quenching	[48]
EuTbL complex with a crown-containing ligand	K^+	MeOH	545Tb-617Eu	0.5–6 eq K^+/Ln		TLC	[58]
$[Eu_{0.47}Tb_{0.53}(L)\ (H_2O)_3]$ MOF	Styrene (vapors)	Air	545Tb-617Eu			PET+FRET quencing	[80]
7:1 Tb:Eu complex with a 3,5-di-carboxy-benzene boronic acid	Ciprofloxacin	H_2O	545Tb-615Eu	0.3–24 μM	90 nM	DAAF	[81]
$[Eu_{0.5}Tb_{0.5}L(H_2O)]Cl$ MOF	F-	H_2O	700Eu-545Tb	0–2 eq F-/Ln	17.7 nM	SQD	[45]
$[Tb_{0.97}Eu_{0.03}(L)\ (H_2O)]$ MOF	F^-		547Tb-617Eu	0–1.9 ppm	96 ppb	SQD	[52]
Hybrid material: luminescent complex [TbL] MOF enclosed in SiO_2 nanoparticles, surface coated with luminescent europium complex Eu (L)	HClO	EtOH	539Tb-607Eu	0–40 μM	0.27 μM	Ligand destruction	[63]
$Eu_{0.05}Tb_{0.95}BTC_{0.9}(L)_{0.1}$ MOF	CrO_4^{2-}		545Tb-618Eu		1μM	IFE+ET	[82]
$Tb_{0.6}Eu_{0.4}$-MOF	Hg^{2+}	H_2O	617Eu-543Tb	0–40 μM	4.83 nM	MMET	[83]
Hybrid material: complex $Eu_4Tb_6(acac)$ on the surface of NZL zeolite	Et_3N, $tBuNH_2$, n-$BuNH_2$, $BnNH_2$, en		545Tb-618Eu			SC	[84]
$[La_{0.88}Eu_{0.02}Tb_{0.1}\ (L)(DMF)_2]_n \cdot H_2O\cdot 0.5DMF$ MOF	HS-	DMF/ H_2O	544Tb-616Eu	0–400 μM		MMET	[57]
	THF	DMF		0–8%		TLC	
	Ag^+	DMF/ H_2O				TLC	

Material	Analyte	Media	Response wavelengths λ1, λ2 (nm)	Linearity range	LOD	Mechanism	Ref.
$Tb_{0.833}Eu_{0.167}(L^{77})_3$ MOF	H_2DPA	H_2O	549Tb-620Eu	2–16 μM	96 nM	MMET+DAAF	[85]
$[Tb_{0.833}Eu_{0.167}BTC]\cdot 6H_2O$ MOF	H_2DPA	H_2O	545Tb-613Eu	0–700 nM	4.55 nM	MMET+DAAF	[86]
$[Eu_{0.06}Tb_{0.04}Gd_{0.9}(L)_{1.5}(H_2O)(DMF)]$ MOF	MnO_4^-	H_2O	613Eu-544Tb	0–0.2 mM	0.02 μM	IFE+FRET	[87]
Ce^{3+}/Tb^{3+} bimetallic guanosine monophosphate complex	alkaline phosphatase	H_2O	552Tb-384Ce	0.2–60 mU mL^{-1}	0.12 mU/mL	Ligand destruction	[42]
$[Eu_{0.1}Tb_{0.9}BTC(H_2O)_2].3DMF.H_2O$ MOF	Coumarin		613Eu-544Tb	1–100 μM	2.3 μM	MMET	[59]
	Caffeine		613Eu-544Tb	1–100 μM		n/a	
	Ln-L materials						
Hybrid material: cerium/terbium adenosine monophosphate impregnated with coumarin solution	Hg^{2+}	H_2O	445–548Tb	0.08–1000 nM	0.03 nM	LIE	[60]
Hybrid material: $CsPbBr_3$@Eu-BTC MOF	Hg^{2+}	H_2O	520L-616Eu	0–1 μM	0.116 nM	PET	[88]
Hybrid material: Eu^{3+}@ Ca-MOF	Hg^{2+}	H_2O	381L-590Eu	0.02–200 μM	2.6 nM	TLC	[89]
$\{[Eu(L)(H_2O)_{1.35}(DMF)_{0.65}]\cdot 1.9DMF\}_n$ MOF	PO_4^{3-}	H_2O	614Eu-368L	0.1–15 μM	52 nM	LIE	[61]
Solution: Eu^{3+} complex with H_2DPA + luminol	PO_4^{3-}	H_2O	615Eu—423L	0.5–50 μM	0.12 μM	QFR (luminol)	[90]
Eu-2′-amino- [1,1′:4′,1″-terphenyl]-3,3″,5,5″-tetracarboxylic acid MOF	F-	H_2O	615Eu-397L	0−5120 μM	11.26 μM	TLC	[91]
Composite material: Tb^{3+} guanosine 5’-disodium@Cu-nanoclusters	alkaline phosphatase	H_2O	545Tb-425L	0.002–2 U mL^{-1}	0.002 U mL^{-1}	LIE	[62]
$Eu_{0.7}Gd_{0.3}(L)_{1.5}(Phen)\cdot H_2O$ complex	Fe^{3+}	H_2O	614Eu-415L	0–60 μM	91 nM	IFE+ET	[72]
Composite material $Eu_{0.7}Gd_{0.3}(tfbdc)_{1.5}(Phen).H_2O$ complex +$FeCl_3$	Ascorbic acid			0–60 μM	0.184 μM	QFR (Fe^{3+})	
Composite material Zn-MOF impregnated with a $TbCl_3$ solution	H_2DPA	H_2O	544Tb-330L	0–1 μM	3.6 nM	SQD+DAAF	[67]

Material	Analyte	Media	Response wavelengths λ1, λ2 (nm)	Linearity range	LOD	Mechanism	Ref.
$EuCl_3$ + sodium polycarylate + pyranine +	H_2DPA	H_2O	615Eu-510L	0–90 μM	10.3 nM	SQD+DAAF	[68]
Eu^{3+}@Zr-MOF UiO-66-$(COOH)_2$-NH_2	H_2DPA	H_2O	621Eu-453L	0–40 μM	25 nM	SQD+DAAF	[69]
$[Tb_2(L^2)_{1.5}$(NMP)2]n MOF	TNP	H_2O	546Tb-400L	0–50 μM	0.11 μM	FRET and PET quenching	[66]
$[Tb_{0.02}Gd_{1.98}L_{1.5}\ (NMP)_2]_n$ MOF		H_2O		0–80 μM	0.41 μM	FRET and PET quenching	
Composite material: Eu MOF in a PMMA matrix	Form-aldehyde	H_2O, Air	453L-616Eu	0.05–1%		Ligand destruction	[64]
$[Dy(L^9)(DMF)_3]_n$ MOF	Nitro-furanose	H_2O	385L-573Dy	0–60 μM	47.6 nM	PET+IFE	[65]
	Furazolidone	H_2O		0–60 μM	48.2 nM	PET+IFE	
$[NMe_2H_2]_2[Tb_9\ (\mu^3\text{-}OH)_8(\mu^2\text{-}OH)_3\ (H_2O)_3(L)_6]\ \cdot 11DMF\cdot 23H_2O$ MOF	Tetra-cycline	H_2O	543Tb-345L	0.14–23 μM	0.18nM	IFE	[71]
Eu^{3+} complex with a luminol (LML) and guanosine 50-monophosphate (GMP) nanoparticles	Tetra-cycline	H_2O	617Eu-430L	0–60 μM	23.2 nM	IFE+ DAAF	[92]
Eu-adenosine-monophosphate@CD	Bisphenol-A	H_2O	429L-623Eu	0–100 μM	20nM	DAAF	[70]
Hybdrid material: Tb@Bi-based MOF	Serotonin	H_2O	545Tb-350L	0–200 μM	0.57 μM	IFE	[93]
Eu^{3+}-functionalized hydrogen-bonded organic framework	CH_3NH_2	H_2O	425L-615Eu	10^{-8}–10^{-2} M	0.87 ppm	IFE	[94]
$[Eu(DMTP\text{-}DC)_{1.5}(H_2O)_3]\cdot$DMF MOF	arginine	H_2O	441L-617Eu	0–250 μM	24.38 μM	PET	[95]
	lysine	H_2O	441L-617Eu	0–250 μM	9.31 μM	PET	
Eu^{3+}@Al-based MOF	β-glucuro nidase	H_2O	450L-617Eu	0.1–50 U L^{-1}	0.03 U L^{-1}	IFE	[96]

Table 3.
"Ratiometric" sensors.

Bisphenol-A, an important industrial reagent in the production of plastics, can act as an effective antenna ligand for europium cations. This underlies the importance of the production of a highly sensitive sensor material based on a composite of carbon quantum dots and europium 5'-adenosine monophosphate [70].

Finally, a response can also be caused by an internal filter effect (IFE). This effect can be the main mechanism, as in the sensor for tetracyclines [71], or manifest itself simultaneously with quenching through ET mechanisms in sensors for Fe^{3+} [72] or for the aforementioned nitrofuranose and furazolidone (**Table 3**) [65].

As in the previous cases, the key stages in studying the mechanism of the "ratiometric" sensory response are the chemical analysis of the reaction product between the sensor and analyte; comparison of the excitation and emission spectra of the sensor and the absorption of the analyte; kinetic studies of excited state lifetimes and quantum chemical modeling. In general, the"ratiometric" response mechanisms usually coincide with those for"turn off" and"turn on" systems. An exception is the mechanism associated with changing the MMET constant ($k^{Ln}{}_{ET}$). This mechanism must be reliably established by kinetic measurements, as well as by studying the response in a mechanical mixture of complexes of two lanthanides.

4. Conclusions

The classification of sensory response mechanisms reported in this review, although perhaps not complete, allows us to successfully classify most of the cited works and make appropriate generalizations, moving from a descriptive style to a debatable one.

The reported examples show the significant progress in the field of lanthanide-based luminescent sensors achieved in recent years. A wide variety of analytes can be qualitatively or quantitatively determined using suitable lanthanide compounds, which requires rational system design. "Turn off" sensors can have a niche application in the analysis of nitroaromatic compounds, while more popular "Turn on" and "ratiometric" materials can be produced using fairly simple strategies: different quenching efficiency of various REE ions by bond vibrations, binding and removal of a quenching fragment by a suitable analyte (e.g., binding Cu^{2+} ions with sulfur-containing ligands), etc.

Progress has touched not only the field of materials design, but also a reliable determination of the sensory response mechanism, which requires several spectroscopic and kinetic measurements, and in some cases quantum chemical calculations of orbital energies. The number of papers containing these studies has been increasing in recent years, and we hope that just such a systematic approach will become the standard in the future works.

Acknowledgements

University of Camerino is gratefully acknowledged.

Conflict of interest

The authors declare no conflict of interest.

Author details

Claudio Pettinari[1]*, Andrei Drozdov[2] and Yuriy Belousov[2]

1 School of Pharmacy, University of Camerino, Camerino, Italy

2 Moscow State University, Moscow, Russia

*Address all correspondence to: claudio.pettinari@unicam.it

References

[1] Bünzli JCG, Eliseeva SV. Basics of lanthanide photophysics. In: Hänninen P, Härmä H, editors. Lanthanide Luminescence. Springer Series on Fluorescence. Vol. 7. Berlin, Heidelberg: Springer; 2010. DOI: 10.1007/4243_2010_3

[2] Belousov YA, Korshunov VM, Metlin MT, Metlina DA, Kiskin MA, Aminev DF, et al. Towards bright dysprosium emitters: Single and combined effects of environmental symmetry, deuteration, and gadolinium dilution. Dyes and Pigments. 2022;**199**:110078. DOI: 10.1016/j.dyepig.2021.110078

[3] Beeby A, Clarkson IM, Dickins RS, Faulkner S, Parker D, Royle L, et al. Non-radiative deactivation of the excited states of europium, terbium and ytterbium complexes by proximate energy-matched OH, NH and CH oscillators: An improved luminescence method for establishing solution hydration states. Journal of the Chemical Society, Perkin Transactions. 1999;**2**(2):493-503. DOI: 10.1039/a808692c

[4] Supkowski RM, Horrocks WDW. On the determination of the number of water molecules, q, coordinated to europium(III) ions in solution from luminescence decay lifetimes. Inorganica Chimica Acta. 2002;**340**:44-48. DOI: 10.1016/S0020-1693(02)01022-8

[5] Kimura T, Kato Y. Luminescence study on hydration states of lanthanide (III)-polyaminopolycarboxylate complexes in aqueous solution. Journal of Alloys and Compounds. 1998;**275–277**: 806-810. DOI: 10.1016/S0925-8388(98)00446-0

[6] Kimura T, Kato Y. Luminescence study on determination of the inner-sphere hydration number of Am(III) and Nd(III). Journal of Alloys and Compounds. 1998;**271–273**:867-871. DOI: 10.1016/S0925-8388(98)00236-9

[7] Chen W, Tang X, Dou W, Wang B, Guo L, Ju Z, et al. The construction of homochiral lanthanide quadruple-stranded helicates with multiresponsive sensing properties toward fluoride anions. Chemistry—A European Journal. 2017;**23**:9804-9811. DOI: 10.1002/chem.201700827

[8] Wang J, Jiang M, Yan L, Peng R, Huangfu M, Guo X, et al. Multifunctional luminescent Eu(III)-based metal-organic framework for sensing methanol and detection and adsorption of Fe(III) ions in aqueous solution. Inorganic Chemistry. 2016;**55**: 12660-12668. DOI: 10.1021/acs.inorgchem.6b01815

[9] Yao Y, Delgado-Rivera L, Samareh Afsari H, Yin L, Thatcher GRJ, Moore TW, et al. Time-gated luminescence detection of enzymatically produced hydrogen sulfide: Design, synthesis, and application of a lanthanide-based probe. Inorganic Chemistry. 2018;**57**:681-688. DOI: 10.1021/acs.inorgchem.7b02533

[10] Aulsebrook ML, Biswas S, Leaver FM, Grace MR, Graham B, Barrios AM, et al. A Luminogenic lanthanide-based probe for the highly selective detection of nanomolar sulfide levels in aqueous samples. Chemical Communications. 2017;**53**:4911-4914. DOI: 10.1039/c7cc01764b

[11] Chen X, Wang Y, Chai R, Xu Y, Li H, Liu B. Luminescent lanthanide-based organic/inorganic hybrid materials for discrimination of glutathione in solution and within hydrogels. ACS Applied

Materials & Interfaces. 2017;**9**:13554-13563. DOI: 10.1021/acsami.7b02679

[12] Li X, Gao J, Rao S, Zheng Y. Development of a selective "on-off-on" nano-sensor based on lanthanide encapsulated carbon dots. Synthetic Metals. 2017;**231**:107-111. DOI: 10.1016/j.synthmet.2017.07.006

[13] Liu B, Shen H, Hao Y, Zhu X, Li S, Huang Y, et al. Lanthanide functionalized metal–organic coordination polymer: Toward novel turn-on fluorescent sensing of amyloid β-peptide. Analytical Chemistry. 2018; **90**:12449-12455. DOI: 10.1021/acs.analchem.8b01546

[14] Yang J, Che J, Jiang X, Fan Y, Gao D, Bi J, et al. A novel turn-on fluorescence probe based on Cu(II) functionalized metal–organic frameworks for visual detection of uric acid. Molecules. 2022; **27**:4803. DOI: 10.3390/molecules27154803

[15] Zhou Z, Li X, Tang Y, Zhang CC, Fu H, Wu N, et al. Oxidative deoximation reaction induced recognition of hypochlorite based on a new fluorescent lanthanide-organic framework. Chemical Engineering Journal. 2018;**351**: 364-370. DOI: 10.1016/j.cej.2018.06.123

[16] Sun T, Hao S, Fan R, Zhang J, Chen W, Zhu K, et al. In situ self-assembled cationic lanthanide metal organic framework membrane sensor for effective MnO4− and ascorbic acid detection. Analytica Chimica Acta. 2021; **1142**:211-220. DOI: 10.1016/j.aca.2020.10.062

[17] Junker AKR, Tropiano M, Faulkner S, Sørensen TJ. Kinetically inert lanthanide complexes as reporter groups for binding of potassium by 18-Crown-6. Inorganic Chemistry. 2016;**55**:12299-12308. DOI: 10.1021/acs.inorgchem.6b02063

[18] Gamonal A, Sun C, Mariano AL, Fernandez-Bartolome E, Guerrero-Sanvicente E, Vlaisavljevich B, et al. Divergent adsorption-dependent luminescence of amino-functionalized lanthanide metal-organic frameworks for highly sensitive NO2Sensors. Journal of Physical Chemistry Letters. 2020;**11**: 3362-3368. DOI: 10.1021/acs.jpclett.0c00457

[19] Liu CL, Zhang RL, Lin CS, Zhou LP, Cai LX, Kong JT, et al. Intraligand charge transfer sensitization on self-assembled europium tetrahedral cage leads to dual-selective luminescent sensing toward anion and cation. Journal of the American Chemical Society. 2017;**139**: 12474-12479. DOI: 10.1021/jacs.7b05157

[20] Li SS, Ye ZN, Xu SS, Zhang YJ, Tao AR, Liu M, et al. Highly luminescent lanthanide CPs based on dinuclear cluster: Crystal structure and sensitive Trp sensor. RSC Advances. 2015;**5**:71961-71967. DOI: 10.1039/c5ra12149c

[21] Ling X, Shi R, Zhang J, Liu D, Weng M, Zhang C, et al. Dual-signal luminescent detection of dopamine by a single type of lanthanide-doped nanoparticles. ACS Sensors. 2018;**3**:1683-1689. DOI: 10.1021/acssensors.8b00368

[22] Niu X, Wang M, Cao R, Zhang M, Liu Z, Liu Z, et al. Ion exchange fabrication of lanthanide functionalized layered double hydroxides microcapsules for rapid and visual detection of anthrax biomarker. Spectrochimica Acta. 2022;**281**:121622. DOI: 10.1016/j.saa.2022.121622

[23] Yang D, Lu L, Feng S, Zhu M. First Ln-MOF as a trifunctional luminescent probe for the efficient sensing of aspartic acid, $Fe^{3+}$and DMSO. Dalton

Transactions. 2020;**49**:7514-7524. DOI: 10.1039/d0dt00938e

[24] Liu B, Huang Y, Shen Q, Zhu X, Hao Y, Qu P, et al. Turn-on fluorescence detection of ciprofloxacin in tablets based on lanthanide coordination polymer nanoparticles. RSC Advances. 2016;**6**:100743-100747. DOI: 10.1039/c6ra20357d

[25] Gupta K, Patra AK. Luminescent Europium(III) "Turn-On" sensor for G-series chemical warfare simulants: A mechanistic investigation. ACS Sensors. 2020;**5**:1268-1272. DOI: 10.1021/acssensors.9b02552

[26] Sahoo J, Arunachalam R, Subramanian PS, Suresh E, Valkonen A, Rissanen K, et al. Coordinatively unsaturated Lanthanide(III) helicates: Luminescence sensors for adenosine monophosphate in aqueous media. Angewandte Chemie. 2016;**128**:9777-9781. DOI: 10.1002/ange.201604093

[27] Liu B, Shen H, Liu D, Hao Y, Zhu X, Shen Q, et al. Citrate/Tb lanthanide coordination polymer nanoparticles: Preparation and sensing of Guanosine-5-Monophosphate. Sensors and Actuators B: Chemical. 2019;**300**:126879. DOI: 10.1016/j.snb.2019.126879

[28] Tang T, Liu M, Chen Z, Wang X, Lai C, Ding L, et al. Highly sensitive luminescent lanthanide metal–organic framework sensor for L-kynurenine. Journal of Rare Earths. 2021;**40**:415-420. DOI: 10.1016/j.jre.2021.02.008

[29] Sergeev AA, Voznesenskiy SS, Petrochenkova NV, Shishov AS, Leonov AA, Emelina TB, et al. Luminescent chemosensors for amines and ammonia based on Eu(III) chelate complexes. In: Proc. SPIE 10176, Asia-Pacific Conference on Fundamental Problems of Opto- and Microelectronics. 14 Dec 2016. p. 1017610. DOI: 10.1117/12.2268131

[30] Voznesenskiy SS, Sergeev AA, Mirochnik AG, Leonov AA, Petrochenkova NV, Shishov AS, et al. Specific features of europium tris-benzoylacetonate sensor response to gaseous ammonia. Sensors and Actuators B: Chemical. 2017;**246**:46-52. DOI: 10.1016/j.snb.2017.02.034

[31] Feng X, Li R, Guo N, Sun Y, Ng SW, Liu X, et al. Two unique hydroxyl bridged lanthanide polymers incorporating mixed carboxylate ligands: Syntheses, structures, luminescence and magnetic property. Inorganica Chimica Acta. 2017;**459**:87-94. DOI: 10.1016/j.ica.2017.01.028

[32] Zheng K, Liu Z, Jiang Y, Guo P, Li H, Zeng C, et al. Ultrahigh luminescence quantum yield lanthanide coordination polymer as a multifunctional sensor. Dalton Transactions. 2018;**47**:17432-17440. DOI: 10.1039/c8dt03832e

[33] Feng X, Feng Y, Guo N, Sun Y, Zhang T, Ma L, et al. Series D-f heteronuclear metal-organic frameworks: Color tunability and luminescent probe with switchable properties. Inorganic Chemistry. 2017;**56**:1713-1721. DOI: 10.1021/acs.inorgchem.6b02851

[34] Sun YL, Feng X, Guo N, Wang LY, Li RF, Bai RF. A novel europium coordination polymer based on mixed carboxylic acid ligands: Synthesis, structure and luminescence. Inorganic Chemistry Communication. 2016;**67**:90-94. DOI: 10.1016/j.inoche.2016.03.006

[35] Hao JN, Yan B. A water-stable lanthanide-functionalized MOF as a highly selective and sensitive fluorescent

probe for Cd^{2+}. Chemical Communications. 2015;**51**:7737-7740. DOI: 10.1039/c5cc01430a

[36] Wang KM, Du L, Ma YL, Zhao JS, Wang Q, Yan T, et al. Multifunctional chemical sensors and luminescent thermometers based on lanthanide metal-organic framework materials. CrystEngComm. 2016;**18**:2690-2700. DOI: 10.1039/c5ce02367j

[37] Li L, Chen Q, Niu Z, Zhou X, Yang T, Huang W. Lanthanide metal-organic frameworks assembled from a fluorene-based ligand: Selective sensing of Pb^{2+} and F^{e3+} ions. Journal of Materials Chemistry C. 2016;**4**:1900-1905. DOI: 10.1039/c5tc04320d

[38] Liu X, Lin H, Xiao Z, Fan W, Huang A, Wang R, et al. Multifunctional lanthanide-organic frameworks for fluorescent sensing, gas separation and catalysis. Dalton Transactions. 2016;**45**: 3743-3749. DOI: 10.1039/c5dt04339e

[39] Yu Y, Li G. Design of Terbium(III)-functionalized covalent organic framework as a selective and sensitive turn-on fluorescent switch for ochratoxin A monitoring. Journal of Hazardous Materials. 2022;**422**:126927. DOI: 10.1016/j.jhazmat.2021.126927

[40] Li YG, Hu JJ, Zhang JL, Liu SJ, Peng Y, Wen HR. Lanthanide-based metal-organic framework materials as bifunctional fluorescence sensors toward acetylacetone and aspartic acid. CrystEngComm. 2022;**24**:2464-2471. DOI: 10.1039/d2ce00174h

[41] Gontcharenko VE, Kiskin MA, Dolzhenko VD, Korshunov VM, Taydakov IV, Belousov YA. Mono- and mixed metal complexes of Eu^{3+}, Gd^{3+}, and Tb^{3+} with a diketone, bearing pyrazole moiety and Chf_2-group: Structure, color tuning, and kinetics of energy transfer between lanthanide ions. Molecules. 2021;**26**:1-16. DOI: 10.3390/molecules26092655

[42] Li J, Gong CC, Li Z, Yao R, Huo P, Deng B, et al. A self-assembly lanthanide nanoparticle for ratiometric fluorescence determination of alkaline phosphatase activity. Journal of Photochemistry and Photobiology A: Chemistry. 2022;**432**: 114054. DOI: 10.1016/j.jphotochem.2022.114054

[43] Li H, Han W, Lv R, Zhai A, Li XL, Gu W, et al. Dual-function mixed-lanthanide metal-organic framework for ratiometric water detection in bioethanol and temperature sensing. Analytical Chemistry. 2019;**91**:2148-2154. DOI: 10.1021/acs.analchem.8b04690

[44] Kornikov AI, Kozlov MI, Lepnev LS, Utochnikova VV. Europium-ytterbium bimetallic complex-based sensor for detecting DMSO impurities in DMSO-D6. Sensors and Actuators B: Chemical. 2022;**370**:132432. DOI: 10.1016/j.snb.2022.132432

[45] Nonat A, Liu T, Jeannin O, Camerel F, Charbonnière LJ. Energy transfer in supramolecular heteronuclear lanthanide dimers and application to fluoride sensing in water. Chemistry—A European Journal. 2018;**24**:3784-3792. DOI: 10.1002/chem.201705532.+

[46] Lakowicz JR. Principles of Fluorescence Spectroscopy. In: Lakowicz JR, editor. Boston, MA: Springer US; 2006

[47] Nakai H, Seo J, Kitagawa K, Goto T, Matsumoto T, Ogo S. An oxygen-sensitive luminescent Dy(III) complex. Dalton Transactions. 2016;**45**:9492-9496. DOI: 10.1039/c6dt01057a

[48] Lehr J, Tropiano M, Beer PD, Faulkner S, Davis JJ. Ratiometric oxygen sensing using lanthanide luminescent

emitting interfaces. Chemical Communications. 2015;**51**:15944-15947. DOI: 10.1039/c5cc05738h

[49] Dunning SG, Nuñez AJ, Moore MD, Steiner A, Lynch VM, Sessler JL, et al. A sensor for trace H_2O detection in D_2O. Chem. 2017;**2**:579-589. DOI: 10.1016/j.chempr.2017.02.010

[50] Gontcharenko VE, Lunev AM, Taydakov IV, Korshunov VM, Drozdov AA, Belousov YA. Luminescent lanthanide-based sensor for H_2O detection in aprotic solvents and D_2O. IEEE Sensors Journal. 2019;**19**:7365-7372. DOI: 10.1109/JSEN.2019.2916498

[51] Chen D, Sun C, Peng Y, Zhang N, Si H, Liu C, et al. Ratiometric fluorescence sensing and colorimetric decoding methanol by a bimetallic lanthanide-organic framework. Sensors and Actuators B: Chemical. 2018;**265**:104-109. DOI: 10.1016/j.snb.2018.03.028

[52] Zeng X, Hu J, Zhang M, Wang F, Wu L, Hou X. Visual detection of fluoride anions using mixed lanthanide metal-organic frameworks with a smartphone. Analytical Chemistry. 2020;**92**: 2097-2102. DOI: 10.1021/acs.analchem.9b04598

[53] Tcelykh L, Kozhevnikova Khudoleeva V, Goloveshkin A, Lepnev L, Popelensky T, Utochnikova V. Sensing of H_2O in D_2O: Is there an easy way? Analyst. 2020;**145**:759-763. DOI: 10.1039/c9an02023c

[54] Li H, Liu B, Xu L, Jiao H. A hetero-MOF-based bifunctional ratiometric fluorescence sensor for PH and water detection. Dalton Transactions. 2021;**50**: 143-150. DOI: 10.1039/D0DT03626A

[55] Xia T, Zhu F, Jiang K, Cui Y, Yang Y, Qian G. A luminescent ratiometric PH sensor based on a nanoscale and biocompatible Eu/Tb-mixed MOF. Dalton Transactions. 2017;**46**:7549-7555. DOI: 10.1039/c7dt01604b

[56] Xia T, Cui Y, Yang Y, Qian G. Highly stable mixed-lanthanide metal-organic frameworks for self-referencing and colorimetric luminescent PH sensing. ChemNanoMat. 2017;**3**:51-57. DOI: 10.1002/cnma.201600331

[57] Wang XY, Yao X, Huang Q, Li YX, An GH, Li GM. Triple-wavelength-region luminescence sensing based on a color-tunable emitting lanthanide metal organic framework. Analytical Chemistry. 2018;**90**:6675-6682. DOI: 10.1021/acs.analchem.8b00494

[58] Yang D, Wang Y, Liu D, Li Z, Li H. Luminescence modulation: Via cation-π interaction in a lanthanide assembly: Implications for potassium detection. Journal of Materials Chemistry C. 2018; **6**:1944-1950. DOI: 10.1039/c7tc04580h

[59] Gao Y, Yu G, Liu K, Wang B. Luminescent mixed-crystal Ln-MOF thin film for the recognition and detection of pharmaceuticals. Sensors and Actuators B: Chemical. 2018;**257**: 931-935. DOI: 10.1016/j.snb.2017.10.180

[60] Zhang Z, Wu Y, He S, Xu Y, Li G, Ye B. Ratiometric fluorescence sensing of mercuric ion based on dye-doped lanthanide coordination polymer particles. Analytica Chimica Acta. 2018;**1014**:85-90. DOI: 10.1016/j.aca.2018.01.065

[61] Cheng Y, Zhang H, Yang B, Wu J, Wang Y, Ding B, et al. Highly efficient fluorescence sensing of phosphate by dual-emissive lanthanide MOFs. Dalton Transactions. 2018;**47**:12273-12283. DOI: 10.1039/C8DT01515E

[62] Li X, Wang X, Guo W, Wang Y, Hua Q, Tang F, et al. Selective detection of alkaline phosphatase activity in

environmental water samples by copper nanoclusters doped lanthanide coordination polymer nanocomposites as the ratiometric fluorescent probe. Biosensors. 2022;**12**. DOI: 10.3390/bios12060372

[63] Ma H, Song B, Wang Y, Cong D, Jiang Y, Yuan J. Dual-emissive nanoarchitecture of lanthanide-complex-modified silica particles for in vivo ratiometric time-gated luminescence imaging of hypochlorous acid. Chemical Science. 2016;**8**:150-159. DOI: 10.1039/C6SC02243J

[64] Wang Y, Zhang G, Zhang F, Chu T, Yang Y. A novel lanthanide MOF thin film: The highly performance self-calibrating luminescent sensor for detecting formaldehyde as an illegal preservative in aquatic product. Sensors and Actuators B: Chemical. 2017;**251**: 667-673. DOI: 10.1016/j.snb.2017.05.063

[65] Wu S, Zhu M, Zhang Y, Kosinova M, Fedin VP, Gao E. A water-stable lanthanide coordination polymer as multicenter platform for ratiometric luminescent sensing antibiotics. Chemistry—A European Journal. 2020; **26**:3137-3144. DOI: 10.1002/chem.201905027

[66] Li L, Cheng J, Liu Z, Song L, You Y, Zhou X, et al. Ratiometric luminescent sensor of picric acid based on the dual-emission mixed-lanthanide coordination polymer. ACS Applied Materials & Interfaces. 2018;**10**:44109-44115. DOI: 10.1021/acsami.8b13719

[67] Zhang D, Zhou Y, Cuan J, Gan N. A lanthanide functionalized MOF hybrid for ratiometric luminescence detection of an anthrax biomarker. CrystEngComm. 2018;**20**:1264-1270. DOI: 10.1039/c7ce01994g

[68] Liu X, Li B, Xu Y, Li Z, Zhang Y, Ding ZJ, et al. A highly selective lanthanide-containing probe for ratiometric luminescence detection of an anthrax biomarker. Dalton Transactions. 2019;**48**:7714-7719. DOI: 10.1039/c9dt01477b

[69] Huo P, Li Z, Yao R, Deng Y, Gong C, Zhang D, et al. Dual-ligand lanthanide metal–organic framework for ratiometric fluorescence detection of the anthrax biomarker dipicolinic acid. Spectrochimica Acta. 2022;**282**:121700. DOI: 10.1016/j.saa.2022.121700

[70] Li Y, Min Q, Wang Y, Zhuang X, Hao X, Tian C, et al. A portable visual coffee ring based on carbon dot sensitized lanthanide complex coordination to detect Bisphenol A in water. RSC Advances. 2022;**12**:7306-7312. DOI: 10.1039/d2ra00039c

[71] Li R, Wang W, El-Sayed ESM, Su K, He P, Yuan D. Ratiometric fluorescence detection of tetracycline antibiotic based on a polynuclear lanthanide metal–organic framework. Sensors and Actuators B: Chemical. 2021;**330**. DOI: 10.1016/j.snb.2020.129314

[72] Yu H, Liu Q, Li J, Su ZM, Li X, Wang X, et al. A dual-emitting mixed-lanthanide MOF with high water-stability for ratiometric fluorescence sensing of $Fe^{3+}$and ascorbic acid. Journal of Materials Chemistry C. 2021;**9**:562-568. DOI: 10.1039/d0tc04781c

[73] Li B, Wang W, Hong Z, El-Sayed ESM, Yuan D. Ratiometric fluorescence detection of trace water in an organic solvent based on bimetallic lanthanide metal-organic frameworks. Chemical Communications. 2019;**55**:6926-6929. DOI: 10.1039/c9cc02324k

[74] Xia D, Li J, Li W, Jiang L, Li G. Lanthanides-based multifunctional luminescent films for ratiometric

humidity sensing, information storage, and colored coating. Journal of Luminescence. 2021;**231**:117784. DOI: 10.1016/j.jlumin.2020.117784

[75] Zhai X, Feng P, Song N, Zhao G, Liu Q, Liu L, et al. Dual-functional ratiometric fluorescent sensor based on mixed-lanthanide metal-organic frameworks for the detection of trace water and temperature. Inorganic Chemistry Frontiers. 2022;**9**:1406-1415. DOI: 10.1039/d2qi00093h

[76] Yu L, Zheng Q, Xiong L, Feng L, Xiao Y. Dual-lanthanide urea metal-organic framework based fluorescent traffic light microsensor for solvent decoding and visual trace water assay. Sensors and Actuators B: Chemical. 2022; **356**:131328. DOI: 10.1016/j.snb.2021.131328

[77] Zhou Q, Dong X, Zhang B, Zhang X, Ou K, Wang Q, et al. Naked-eye sensing and target-guiding treatment of bacterial infection using PH-tunable multicolor luminescent lanthanide-based hydrogel. Journal of Colloid and Interface Science. 2022;**610**:731-740. DOI: 10.1016/j.jcis.2021.11.121

[78] Qin SJ, Qu XL, Yan B. A self-calibrating bimetallic lanthanide metal-organic luminescent sensor integrated with logic gate operation for detecting N-methylformamide. Inorganic Chemistry Frontiers. 2018;**5**:2971-2977. DOI: 10.1039/c8qi00958a

[79] Zhou J, Li H, Zhang H, Li H, Shi W, Cheng P. A bimetallic lanthanide metal-organic material as a self-calibrating color-gradient luminescent sensor. Advanced Materials. 2015;**27**:7072-7077. DOI: 10.1002/adma.201502760

[80] Su Y, Yu J, Li Y, Phua SFZ, Liu G, Lim WQ, et al. Versatile bimetallic lanthanide metal-organic frameworks for tunable emission and efficient fluorescence sensing. Commun. Chem. 2018;**1**. DOI: 10.1038/s42004-018-0016-0

[81] Sun Y, Dramou P, Song Z, Zheng L, Zhang X, Ni X, et al. Lanthanide metal doped organic gel as ratiometric fluorescence probe for selective monitoring of ciprofloxacin. Microchemical Journal. 2022;**179**:107476. DOI: 10.1016/j.microc.2022.107476

[82] Wu RZ, Yang X, Zhang LW, Zhou PP. Luminescent lanthanide metal-organic frameworks for chemical sensing and toxic anion detection. Dalton Transactions. 2017;**46**:9859-9867. DOI: 10.1039/c7dt01790a

[83] Wang X, Jiang Z, Yang C, Zhen S, Huang C, Li Y. Facile synthesis of binary two-dimensional lanthanide metal-organic framework nanosheets for ratiometric fluorescence detection of mercury ions. Journal of Hazardous Materials. 2022;**423**:126978. DOI: 10.1016/j.jhazmat.2021.126978

[84] Li P, Li H. Amine vapor responsive lanthanide complex entrapment: Control of the ligand-to-metal and metal-to-metal energy transfer. Journal of Materials Chemistry C. 2016;**4**:2165-2169. DOI: 10.1039/c5tc04377h

[85] Gao N, Zhang Y, Huang P, Xiang Z, Wu FY, Mao L. Perturbing Tandem energy transfer in luminescent heterobinuclear lanthanide coordination polymer nanoparticles enables real-time monitoring of release of the anthrax biomarker from bacterial spores. Analytical Chemistry. 2018;**90**:7004-7011. DOI: 10.1021/acs.analchem.8b01365

[86] Zhang Y, Li B, Ma H, Zhang L, Jiang H, Song H, et al. A nanoscaled lanthanide

metal-organic framework as a colorimetric fluorescence sensor for dipicolinic acid based on modulating energy transfer. Journal of Materials Chemistry C. 2016;**4**:7294-7301. DOI: 10.1039/c6tc01022a

[87] Dong ZP, Zhao F, Zhang L, Liu ZL, Wang YQ. A white-light-emitting lanthanide metal-organic framework for luminescence turn-off sensing of MnO_4- and turn-on sensing of folic acid and construction of a "Turn-on plus" system. New Journal of Chemistry. 2020;**44**: 10239-10249. DOI: 10.1039/d0nj02145h

[88] Shu Y, Ye Q, Dai T, Guan J, Ji Z, Xu Q, et al. Incorporation of perovskite nanocrystals into lanthanide metal-organic frameworks with enhanced stability for ratiometric and visual sensing of mercury in aqueous solution. Journal of Hazardous Materials. 2022; **430**:128360. DOI: 10.1016/j.jhazmat.2022.128360

[89] Hao Guo NW, Peng L, Chen Y, Liu Y, Li C, Zhang H, et al. A novel ratiometric fluorescence sensor based on lanthanide-functionalized MOF for Hg2+ detection. Talanta. 2022;**250**:123710. DOI: 10.1016/j.talanta.2022.123710

[90] Wu H, Ling Y, Ju S, Chen Y, Xu M, Tang Y. A smartphone-integrated light-up lanthanide fluorescent probe for the visual and ratiometric detection of total phosphorus in human urine and environmental water samples. Spectrochimica Acta. 2022;**2022**:279. DOI: 10.1016/j.saa.2022.121360

[91] Yu K, Wang Q, Xiang W, Li Z, He Y, Zhao D. Amino-functionalized single-lanthanide metal-organic framework as a ratiometric fluorescent sensor for quantitative visual detection of fluoride ions. Inorganic Chemistry. 2022;**61**: 13627-13636. DOI: 10.1021/acs.inorgchem.2c02533

[92] Yin S, Tong C. Lanthanide coordination polymer nanoparticles as a ratiometric fluorescence sensor for real-time and visual detection of tetracycline by a smartphone and test paper based on the analyte-triggered antenna effect and inner filter effect. Analytica Chimica Acta. 2022;**1206**:339809. DOI: 10.1016/j.aca.2022.339809

[93] Song L, Tian F, Liu Z. Lanthanide doped metal-organic frameworks as a ratiometric fluorescence biosensor for visual and ultrasensitive detection of serotonin. Journal of Solid State Chemistry. 2022;**312**:123231. DOI: 10.1016/j.jssc.2022.123231

[94] Liu Y, Xu X, Lu H, Yan B. Dual-emission ratiometric fluorescent probe-based lanthanide-functionalized hydrogen-bonded organic framework for the visual detection of methylamine. Journal of Materials Chemistry C. 2022; **10**:1212-1219. DOI: 10.1039/d1tc04613f

[95] Hu JJ, Li YG, Wen HR, Liu SJ, Peng Y, Liu CM. Stable lanthanide metal-organic frameworks with ratiometric fluorescence sensing for amino acids and tunable proton conduction and magnetic properties. Inorganic Chemistry. 2022; **61**:6819-6828. DOI: 10.1021/acs.inorgchem.2c00121

[96] Li S, Yu L, Xiong L, Xiao Y. Ratiometric fluorescence and chromaticity dual-readout assay for β-glucuronidase activity based on luminescent lanthanide metal-organic framework. Sensors and Actuators B: Chemical. 2022;**355**:131282. DOI: 10.1016/j.snb.2021.131282

Chapter 5

Prediction of Solubility and Miscibility Parameters of Bismuth-Arsenic Complex and Amorphous Mineral Compounds Using Molecular Dynamics Simulation

Francisco Adrián De la Torre-Martínez, Efren Delgado, María Dolores Josefina Rodríguez Rosales, Hiram Medrano-Roldán, Javier López-Miranda and Damián Reyes-Jáquez

Abstract

Bismuth is one of the most difficult impurities to remove in mining concentrates and low concentrations generate problems in silver and copper refineries. Therefore, financial penalties are established when concentrations exceed 0.05%. Some researchers had used arsenic to remove bismuth with results of up to 52% of extraction. Unfortunately, this mechanism is not yet fully understood. The objective of this research was to obtain the solubility parameters of amorphous mineral compounds, including bismuth-based compounds, through computational simulation using molecular dynamics. The composition of the mineral sample was determined by X-ray diffraction and the crystalline species were obtained and modeled using Materials Studio software. The nanostructures were optimized by an energy minimization methodology using the Broyden-Fletcher-Goldfarb-Shanno algorithm and were validated using the figure of merit equation and density. Simulations were performed using the Universal Force Field at constant pressure and temperature. The results of the minerals identified in the sample were compared with arsenic trioxide, indicating miscibility between As_2O_3 and Bi_2O_3, possible miscibility with 10 other minerals, and immiscibility with the rest. The results indicate that As_2O_3 can be successfully used for the removal of Bi_2O_3 without a negative effect on the recovery of other minerals of higher commercial value.

Keywords: molecular dynamics, solubility parameter, miscibility, amorphous mineral compounds, bismuth, arsenic trioxide

1. Introduction

In many metals' extraction processes, the ore extracted from the mines usually undergoes a particle reduction or purification process, where high amounts of amorphous phases can be generated [1]. The presence of impurities (non-desired compounds) affects properties, such as electrical conductivity and ductility [2]. Bismuth (Bi), a very difficult element to remove from copper minerals, is practically insoluble in such minerals because of its melting point of 271°C, compared to 1083°C for copper, and tends to localize at the edges of copper crystals, causing rod cracking and poor drawability [3]. Regularly, high commercial value metals with bismuth concentrations higher than those established by the buyer result in penalties, thus decreasing the final price, so the elimination of such impurities in the extraction processes has become a key problem faced mainly by mining companies [4]. A negative setback is the lack of detailed research on its removal.

Previously Xiao et al. [1] developed a method using arsenic trioxide to form a complex that precipitates and which can be later removed. Xiao and Zheng found that As(III) ion as As_2O_3 played a significant role in the removal of antimony (Sb) and Bi impurities from copper electrolytes [1, 5, 6]. When As(III) is added to the electrolyte, the removal rate of antimony and bismuth increases significantly [5, 7], reaching up to 52% of extraction effectiveness. However, despite being an effective method, this mechanism is not yet fully understood, so it is desirable to further evaluate it and a key consideration is to know the miscibility between the compounds of interest.

One technique that allows a detailed and profound understanding of this mechanism is by using a computational simulation by molecular dynamics (MD), where a molecular structure and its behavior can be represented and analyzed over time, by considering intramolecular and intermolecular forces, such as van der Waals forces, electrostatic forces, covalent and non-covalent bonds, among others. From the simulation, different physicochemical properties of the system can be calculated, such as the solubility parameter of the substance, which, according to the principle of similarity, if two substances have similar values it indicates that they are miscible, hence it could predict whether two substances are miscible or not [8, 9].

Therefore, in this research, a computational simulation by molecular dynamics was used through Materials Studio 8.0 software to obtain the solubility parameters of simulated amorphous cells of arsenic trioxide (As_2O_3) and different minerals that compose a sample of mining concentrate, among them bismuth oxide (Bi_2O_3), thus indicating the miscibility and affinity between them for the formation of bismuth-arsenic complexes.

2. Materials and methods

The mineral sample was obtained from a local mine located in the Sierra Madre Occidental region of Durango, Mexico, and it was grinded and sieved using a 200 mesh.

2.1 X-ray diffraction

X-ray diffraction (XRD) of the powdered mineral sample was carried out on a Mini Flex 300 equipment in a 2θ angular range between 5° and 90° at a step size of 0.02°.

DOI: http://dx.doi.org/10.5772/intechopen.106316

The obtained diffraction pattern was refined using the Rietveld method and the crystalline composition was determined. The crystal structures were also obtained from the open database of crystallography (COD) [10–14].

2.2 Molecular dynamics simulation

BIOVA Materials Studio software version 8.0 (Accelrys, San Diego, CA) was used to perform the molecular dynamics simulations. The crystals were created using the latency data previously obtained from the XRD technique, from which the amorphous cells were obtained with the amorphous cell module; 10 cubic boxes were created, containing 100 repetitions of each of the molecules at 298 K. Geometric optimization was performed using the FORCITE module using the Universal Force Field (UFF) and the Quasi-Newton Broyden-Fletcher-Goldfarb-Shanno (BFGS) algorithm using the sum of atoms method for nonderivative van der Waals and electrostatic interactions, which were truncated using a group distance less than half the box length. The Andersen thermostat and barostat were used to maintain temperature and pressure, respectively. The box with the lowest energy was used to perform dynamics using constant number of molecules, constant pressure, and constant temperature (NPT) of 1 atm and 298 K, respectively, and for a time of 200 ps with a time step of 1 fs. Finally, the cohesive energy densities and the solubility parameters were obtained.

3. Results

The XRD results are shown in **Figure 1**, where silicon oxide or quartz (Si_2O) predominates since the sample was not previously processed. Silver and gold, among others, were also found. Bismuth was detected as an element of interest, since it is considered an impurity and is the object of study of the research, due to the economic interests of the mining industry.

The Mini Flex 300 comes equipped with a program and a crystal database that provides crystallographic data, such as lattice parameters, crystal type, and space

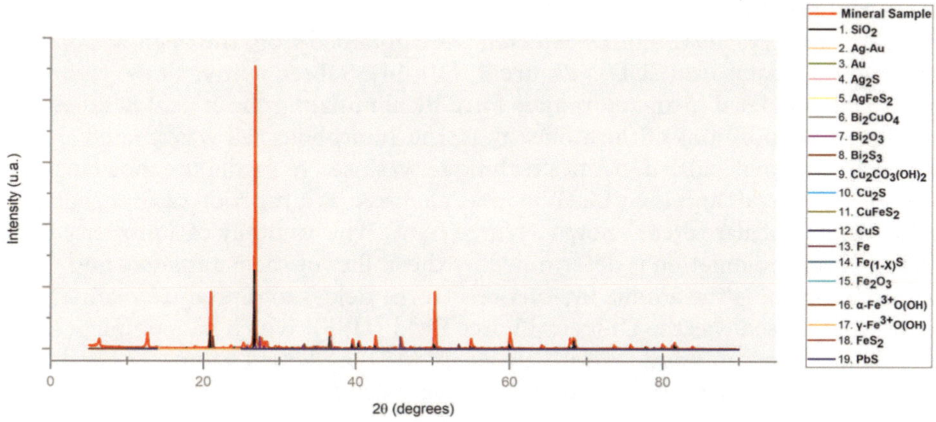

Figure 1.
Diffractogram of the mineral sample and its crystalline composition.

Name	Formula	Space group	Crystal type	Lattice parameters				
				a (Å)	b (Å)	c (Å)	β (°)	γ (°)
Silver	Ag	225: Fm-3m	Cubic	4.09	4.09	4.09	90.00	90.00
Argentite	Ag_2S	229: Im-3m	Cubic	4.88	4.88	4.88	90.00	90.00
Lenaite	$AgFeS_2$	122: I-42d	Tetragonal	5.94	5.94	10.17	90.00	90.00
Gold	Au	194: P63/mmc	Cubic	2.35	2.35	4.00	90.00	120.00
Kusachiita	Bi_2CuO_4	130: P4/ncc	Tetragonal	8.56	8.56	5.79	90.00	90.00
Bismite	Bi_2O_3	14: P21/c	Monoclinic	5.85	8.17	7.51	112.58	90.00
Bismuthine	Bi_2S_3	62: Pnma	Orthorhombic	11.04	4.25	10.33	90.00	90.00
Malachite	$Cu_2(CO_3)(OH)_2$	14: P121	Monoclinic	9.41	10.85	2.92	95.90	90.00
Chalcocite	Cu_2S	194: P63/mmc	Monoclinic	3.57	3.57	6.29	90.00	120.00
Chalcopyrite	$CuFeS_2$	122: I-42d	Tetragonal	5.28	5.28	10.41	90.00	90.00
Covellite	CuS	194: P63/mmc	Hexagonal	4.00	4.00	16.00	90.00	120.00
Iron	Fe	225: Fm-3m	Cubic	2.83	2.83	2.83	90.00	90.00
Pyrrhotite	$Fe_{(1-x)}S$	147: P-3	Monoclinic	6.14	6.14	57.40	90.00	120.00
Hematite	Fe_2O_3	167: R-3c	Hexagonal	5.02	5.02	13.78	90.00	120.00
Goethite	$Fe^{3+}O(OH)$	62: Pbnm	Orthorhombic	4.50	9.80	2.90	90.00	90.00
Marcasite	FeS_2	58: Pnnm	Orthorhombic	4.52	5.43	3.38	90.00	90.00
Galena	PbS	225: Fm-3m	Cubic	6.20	6.20	6.20	90.00	90.00
Quartz	SiO_2	154: P3221	Hexagonal	4.91	4.91	5.40	90.00	120.00
Lepidocrocite	ϒ-FeO(OH)	63: Cmcm	Orthorhombic	3.13	12.37	3.98	90.00	90.00

Table 1.
Crystallographic data of the mining sample.

group, among others (**Table 1**). Subsequently, the composition of each mineral in the sample was determined, as shown in **Figure 2**.

These data obtained from XRD were supplied to the Materials Studio 8.0 program for the creation of the *in silico* crystals (**Figure 3**).

Initially, the crystal structures detected were obtained from the open crystallography database (COD) (**Figure 3**) [10–14]. Subsequently, all the created crystals were modeled in an amorphous form by eliminating the crystal parameters and keeping the positions of the atoms; thus, the amorphous cell was created (**Figure 4**). The molecular dynamics technique was used to model the molecular structure of minerals and their behavior was analyzed as a function of time, considering the intramolecular forces known as force fields. The accuracy of a properly balanced molecular simulation is determined by the ability of the parameters and equations describing the atomic interactions (force fields) to reproduce reality [15]. The force field used was the Universal Force Field (UFF), which has complete coverage of the periodic table and is relatively accurate in predicting geometries and conformational energy differences of metal complexes and is expressed by Eq. (1) [16, 17].

$$E_{pot} = E_R + E_\theta + E_\phi + E_\omega + E_{vdW} + E_{el} \tag{1}$$

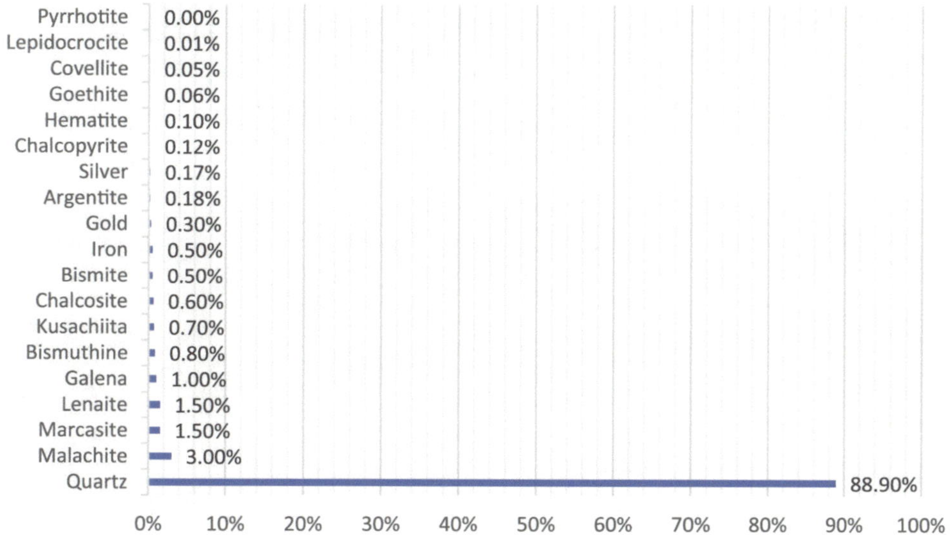

Figure 2.
Qualitative analysis of the mining sample (weight percentage of the crystalline components).

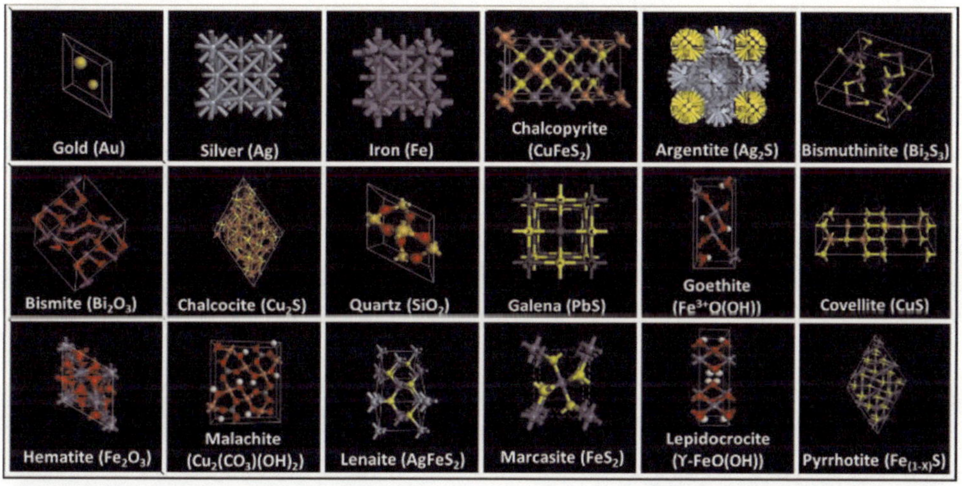

Figure 3.
Crystals created by simulation in Materials Studio.

where E_R = bond-stretching energy, E_θ = valence angle bending energy, E_Φ = dihedral torsion energy, E_ω = inversion energy, E_{vdW} = van der Waals interaction energy, and E_{el} = electrostatic energy.

The chosen force field was developed to be used with all the elements of the periodic table to obtain appropriate parameters. In this force field, the first three terms parameterize the short-range bound interactions, the fourth is the energy required to transform a molecule from one spatial form to another, while the last two terms parameterize the unbound inter and intramolecular interactions [16].

Normally the initial structures that are created in the simulators have much higher energies than a real structure would have, for this reason, algorithms are used to

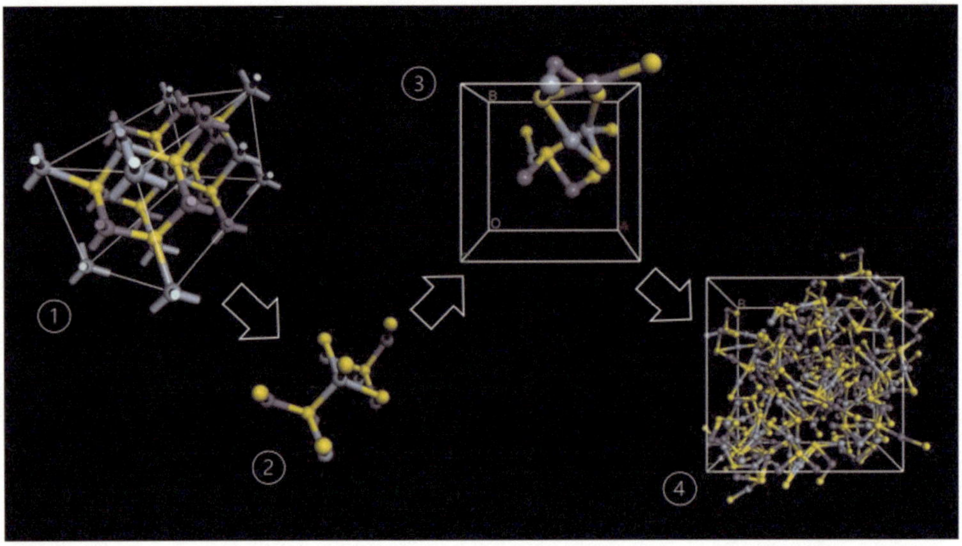

Figure 4.
Creation of the amorphous cell from (1) $AgFeS_2$ crystal, (2) $AgFeS_2$ molecular structure without lattice parameters, (3) AgFeS2 unit amorphous cell, (4) amorphous cell with 100 units of $AgFeS_2$ repetitions, (5) $AgFeS_2$ unit amorphous cell with 100 units of $AgFeS_2$ repetitions, and (6) $AgFeS_2$ unit amorphous cell with 100 units of $AgFeS_2$ repetitions.

calculate the positions and forces, to minimize them and make them more realistic. The minimum points on the potential energy surface corresponding to the steady states of the system were then analyzed and a geometric optimization using the Quasi-Newton method with the Broyden-Fletcher-Goldfarb-Shanno algorithm (BFGS) was performed to achieve the crystal relaxation and improve its stability and efficiency [18]. BFGS accumulates information about the Hessian matrix and the shape of the enthalpy surface around the minimum [19].

Figure 5 shows (a) the amorphous cell of arsenic trioxide in its initial form where the total energy is above 12 million kcal/mol, and, in image (b), the same cell after

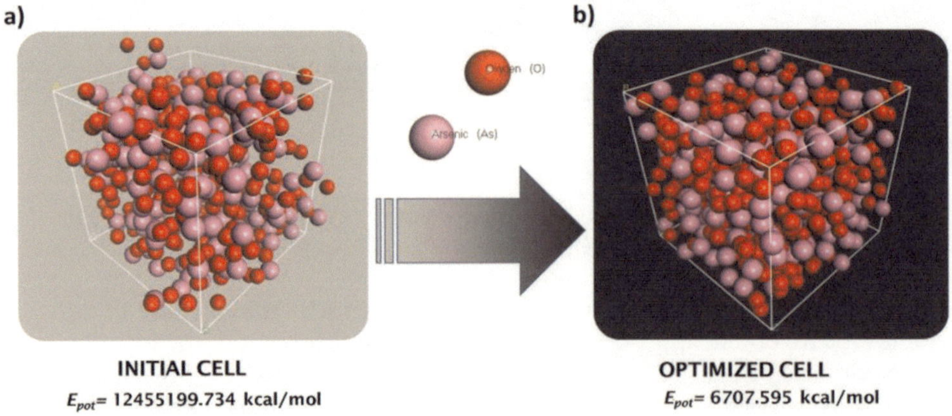

Figure 5.
Energy-minimized structure of As_2O_3 at (a) its initial configuration at 298 K, and (b) the end of the geometry optimization.

optimization at 298 K where the energy is 6707.6 kcal/mol, indicating that it is a structure closer to its minimum energy state, and, therefore, more stable and likely to be found in reality.

The rationality of the simulation system was verified by comparing the simulated density with the experimental reference density from the Mining Handbook published by the Mineralogical Society of America [20], as shown in **Table 2**, since density is one of the most basic physical properties. Gupta mentions that close agreement between simulated and experimental density values means effective optimization of intermolecular interactions and hence accurate prediction [21]. All density values remained within the correct ranges, except for Kusachiite and Argentite, which showed a discrepancy of 1.16 and 1.67%, respectively.

The density calculated in the simulation is obtained using Eq. (2) [22].

$$\rho_{calc} = \frac{N_{at} \bullet M}{V_{cel} \bullet N_A} \tag{2}$$

Name	Formula	Density g/cm^3	
		Simulated	Experimental
Silver	Ag	10.50	10.1–11.1
Argentite	Ag_2S	7.08	7.2–7.4
Lenaite	$AgFeS_2$	4.60	4.6
Claudetite	As_2O_3	4.17	3.9–4.2
Arsenolite	As_2O_3	3.87	3.86–3.88
Gold	Au	18.40	16–19.3
Kusachiita	Bi_2CuO_4	8.60	8.5
Bismite	Bi_2O_3	9.37	8.5–9.5
Bismuthine	Bi_2S_3	6.81	6.8 7.2
Malachite	$Cu_2(CO_3)(OH)_2$	4.00	3.6–4
Chalcocite	Cu_2S	5.74	5.5–5.8
Chalcopyrite	$CuFeS_2$	4.19	4.1–4.3
Covellite	CuS	4.63	4.6–4.8
Iron	Fe	7.87	7.3–7.9
Pyrrhotite	$Fe_{(1-x)}S$	4.62	4.6–4.7
Hematite	Fe_2O_3	5.30	5.2–5.3
Goethite	$Fe^{3+}O(OH)$	4.27	3.3–4.3
Marcasite	FeS_2	4.88	4.8–4.9
Galena	PbS	7.37	7.2–7.6
Quartz	SiO_2	2.65	2.6–2.7
Lepidocrocite	ϒ-FeO(OH)	4.00	4

Table 2.
Comparison of simulated and experimental densities.

Name	Formula	FoM
Arsenolite	As_2O_3	0.94
Bismite	Bi_2O_3	0.91
Bismuthine	Bi_2S_3	0.91
Galena	PbS	0.85
Lenaite	$AgFeS_2$	0.98
Quartz	SiO_2	0.89
Chalcopyrite	$CuFeS_2$	0.98
Covellite	CuS	0.92
Gold	Au	0.98
Kusachiita	Bi_2CuO_4	0.96
Marcasite	FeS_2	0.98
Hematite	Fe_2O_3	0.97
Lepidocrocite	ϒ-FeO(OH)	0.95
Goethite	$Fe^{3+}O(OH)$	0.96
Iron	Fe	0.99

Table 3.
Comparison of results of simulated and experimental XRD patterns.

where P_{calc} = calculated density, N_{at} = number of atoms per cell, M = atomic mass, V_{cel} = volume of the cell, and N_A = Avogadro's number = 6.022 140 857(74) $\times$ 10^{23} 1/mol.

For validation, the experimental and simulated XRD patterns were used, and, using Eq. (3), which corresponds to the figure of merit (FoM) [23], the similarity between them was evaluated using the total number of coincident peaks, as well as the intensities and values at their highest peaks (**Table 3**).

$$FoM = \sqrt{\frac{FoM_{db} \bullet (w_\theta \bullet FoM_\theta + w_I \bullet FoM_I + w_{ph} \bullet FoM_{ph})}{w_\theta + w_I + w_{ph}}} \tag{3}$$

where FoM_{db} is the contribution due to the associated database peak intensities and its percentage, FoM_θ coming from 2θ is the difference between the experimental database peaks, FoM_I is the difference between the experimental database peak intensities, FoM_{ph} is the contribution due to the associated experimental peak intensities and its percentage, and w_θ, w_I, and w_{ph} are weighting factors.

The patterns of the simulated crystals coincided consistently with the real ones in the database, those values greater than 0.8 are considered acceptable for identification. The patterns of As_2O_3 and Bi_2O_3 are shown in **Figure 6**.

Hildebrand [24] introduced the concept of solubility parameter (δ_{hil}), defined as the square root of the cohesive energy density (CED) (Eq. (3)), which is the energy required to break intermolecular bonds per unit volume. Those compounds with similar solubility parameters are miscible in most proportions, while values with a greater difference are considered immiscible. Greenhalgh et al. [25] and Forster et al. [26] mention that those values with differences less than 2 would be miscible, values

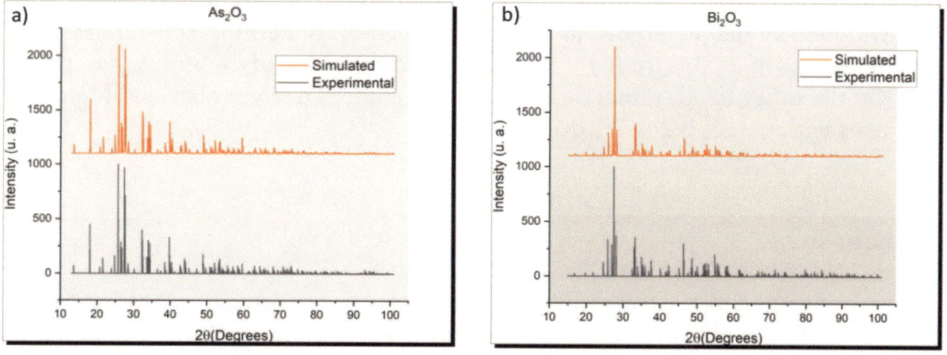

Figure 6.
Comparison of simulated XRD pattern with the experimental one for (a) As_2O_3 and (b) Bi_2O_3.

Name	Formula	$\delta_{Hildebrand}$	$(\lvert\delta_{Arsenolite}-\delta_{Mineral}\rvert)$	Miscibility with As_2O_3
Arsenolite	As_2O_3	22.65	—	—
Bismite	Bi_2O_3	24.33	1.68	Miscible
Marcasite	FeS_2	20.49	2.15	Possibly miscible
Quartz	SiO_2	24.91	2.26	Possibly miscible
Bismuthine	Bi_2S_3	25.86	3.21	Possibly miscible
Galena	PbS	27.64	4.99	Possibly miscible
Malachite	$Cu_2(CO_3)(OH)_2$	17.23	5.42	Possibly miscible
Lenaite	$AgFeS_2$	16.96	5.69	Possibly miscible
Kusachiita	Bi_2CuO_4	16.42	6.23	Possibly miscible
Argentite	Ag_2S	16.21	6.44	Possibly miscible
Covellite	CuS	15.90	6.75	Possibly miscible
Pyrrhotite	$Fe_{(1-X)}S$	15.60	7.05	Possibly miscible
Chalcopyrite	$CuFeS_2$	15.53	7.12	Immiscible
Lepidocrocite	ϒ-FeO(OH)	14.78	7.87	Immiscible
Goethite	$Fe^{3+}O(OH)$	14.78	7.87	Immiscible
Chalcocite	Cu_2S	12.82	9.83	Immiscible
Hematite	Fe_2O_3	12.55	10.09	Immiscible
Silver	Ag	10.31	12.34	Immiscible
Gold	Au	10.00	12.65	Immiscible
Iron	Fe	6.95	15.70	Immiscible

Table 4.
Solubility parameters in $MPa^{0.5}$ from the simulation of amorphous mineral cells compared to As_2O_3.

between 2 and 7 would possibly be miscible and, on the other hand, values greater than 7 would be immiscible. The solubility parameter is calculated by Eq. (4) [27].

$$\delta_{hil} = \sqrt{CED} = \sqrt{\frac{E_{coh}}{V}} \tag{4}$$

where E_{coh} is cohesive energy and V is the molar volume of the substance.

Table 4 shows the difference in crystal parameters concerning arsenic trioxide, obtaining values of 24.33 $(J/cm^3)^{0.5}$ and 22.65 $(J/cm^3)^{0.5}$ for Bi_2O_3 and As_2O_3, respectively. On the other hand, minerals such as iron, gold, and silver obtained higher differences with values below 11.00 $(J/cm^3)^{0.5}$.

4. Conclusions

The composition of the mineral sample from the Avino mine consisted of 19 identified crystals, with three of them containing bismuth in the form of B_iO_3, Bi_2S_3, and Bi_2CuO_4: representing 2% of the total composition of the sample, while also contributing to 1.65% of bismuth weight, exceeding the allowable limit of 0.05% for bismuth by far. The structures of the simulated minerals were effectively corroborated to match the real structures by comparing the diffraction patterns obtaining similarities higher than 9. The solubility parameters indicate an affinity in the miscibility between As_2O_3 and Bi_2O_3, possible miscibility with 10 other minerals, and an immiscibility with the rest of the minerals in which gold, silver, and copper stand out. The results indicate that As_2O_3 can be used for the removal of Bi_2O_3 at ambient pressure and temperature conditions without having a negative effect on the recovery of other minerals of higher commercial value. The mining industry could benefit from the results obtained from this research since the economic penalties because of the presence of bismuth in concentrates might be reduced by using arsenic trioxide. The molecular simulation could serve as a resourceful tool in metal recovery processing by avoiding time investment and the use of reagents.

Acknowledgements

This work was supported in part by the USDA National Institute of Food and Agriculture, Hatch grant (1010849) “Food Bioengineering Technology of Agro-industrial Products.”

Conflict of interest

The authors declare no conflict of interest.

Author details

Francisco Adrián De la Torre-Martínez[1], Efren Delgado[2], María Dolores Josefina Rodríguez Rosales[1], Hiram Medrano-Roldán[1], Javier López-Miranda[1] and Damián Reyes-Jáquez[1]*

1 TecNM—Durango Institute of Technology, Durango, México

2 Food Science and Technology, Department of Family and Consumer Sciences, New Mexico State University, Las Cruces, New Mexico, USA

*Address all correspondence to: damian.reyes@itdurango.edu.mx

References

[1] Xiao F, Mao J, Cao D, Shen X, Volinsky AA. The role of trivalent arsenic in removal of antimony and bismuth impurities from copper electrolytes. Hydrometallurgy. 2012;**125–126**:76-80. DOI: 10.1016/j.hydromet.2012.05.011

[2] Zeng W, Hui H, Xiao R, Yang J, Liu S, Lin W, et al. Study of in-situ precipitation of arsenic bearing crystalline particles during the process of copper electrorefining. Hydrometallurgy. 2021;**199**:105546. DOI: 10.1016/j.hydromet.2020.105546

[3] Davenport WG, King M, Schlesinger M, Biswas AK. Extractive Metallurgy of Copper. Pergamon; 2002

[4] Vargas-Rubio KI, Medrano-Roldan H, Reyes-Jáquez D. Comparative study of chemical process and biotechnological process for the removal of bismuth from mining concentrates. Revista Mexicana de Ingeniería Química. 2021;**20**(3):1-8

[5] Fa-xin X, Ya-jie Z, Yong W. Purification mechanism of copper electrolyte by As(III). Transactions of Nonferrous Metals Society of China. 2008;**18**(5):1275-1279. DOI: 10.1016/S1003-6326(08)60216-2

[6] Ya-jie Z et al. Industrial experiment of copper electrolyte purification by copper arsenite. Journal of Central South University of Technology. 2008;**15**:204-208. DOI: 10.1007/s11771-008-0039-2

[7] Andrew A. Removal of antimony and bismuth from copper electrorefining electrolyte by two proprietary solvent extraction extractants [masters theses]. 2019. p. 7877. Available from: https://scholarsmine.mst.edu/masters_theses/7877

[8] Salehi HS, Ramdin M, Moultos OA, Vlugt TJH. Fluid phase equilibria computing solubility parameters of deep eutectic solvents from molecular dynamics simulations. Fluid Phase Equilibria. 2019;**497**:10-18. DOI: 10.1016/j.fluid.2019.05.022

[9] Salehi HS, Ramdin M, Moultos OA, Vlugt TJH. Computing solubility parameters of deep eutectic solvents from molecular dynamics simulations. Fluid Phase Equilibria. 2019;**497**:10-18. DOI: 10.1016/j.fluid.2019.05.022

[10] Vaitkus A, Merkys A, Gražulis S. Validation of the crystallography open database using the crystallographic information framework. Journal of Applied Crystallography. 2021;**54**(2):661-672. DOI: 10.1107/S1600576720016532

[11] Quirós M, Gražulis S, Girdzijauskaitė S, Merkys A, Vaitkus A. Using SMILES strings for the description of chemical connectivity in the Crystallography Open Database. Journal of Cheminformatics. 2018;**10**(1):23

[12] Gražulis S, Merkys A, Vaitkus A, Okulič-Kazarinas M. Computing stoichiometric molecular composition from crystal structures. Journal of Applied Crystallography. 2015;**48**(1):85-91. DOI: 10.1107/S1600576714025904

[13] Gražulis S et al. Crystallography Open Database (COD): An open-access collection of crystal structures and platform for world-wide collaboration. Nucleic Acids Research. 2012;**40**(D1):D420-D427. DOI: 10.1093/nar/gkr900

[14] Gražulis S et al. Crystallography Open Database—An open-access collection of crystal structures. Journal of

Applied Crystallography. 2009;**42**(4): 726-729. DOI: 10.1107/S0021889809016690

[15] Martin MG. Comparison of the AMBER, CHARMM, COMPASS, GROMOS, OPLS, TraPPE and UFF force fields for prediction of vapor-liquid coexistence curves and liquid densities. Fluid Phase Equilibria. 2006;**248**(1): 50-55. DOI: 10.1016/j.fluid.2006.07.014

[16] Rappe AK, Colwell KS, Casewit CJ. Application of a universal force field to metal complexes. Inorganic Chemistry. 1993;**32**(16):3438-3450. DOI: 10.1021/ic00068a012

[17] Arab B, Shokuhfar A. Molecular dynamics simulation of cross-linked epoxy polymers: The effect of force field on the estimation of properties. Journal of Nano- and Electronic Physics. 2013; **5**(1):1-5

[18] Pfrommer BG, Côté M, Louie SG, Cohen ML. Relaxation of crystals with the Quasi-Newton method. Journal of Computational Physics. 1997;**131**(1): 233-240. DOI: 10.1006/jcph.1996.5612

[19] Vargas Rubio KI, Medrano Roldán H, Reyes Jáquez D. Molecular dynamics simulation of a nanocluster obtained from the mining industry. Acta Universitaria. 2021;**31**:1-10

[20] Anthony JW, Bideaux RA, Bladh KW, Nichols MC. Handbook of Mineralogy, Elements, Sulfides and Sulfosalts. Mineral Data Publishing; 1990

[21] Gupta J, Nunes C, Vyas S, Jonnalagadda S. Prediction of solubility parameters and miscibility of pharmaceutical compounds by molecular dynamics simulations. Journal of Physical Chemistry B. 2011;**115**(9): 2014-2023. DOI: 10.1021/jp108540n

[22] BIOVIA. BIOVA Materials Studio. Dassault Systèmes [Online]. 2019. Available from: http://accelrys.com/products/collaborative-science/biovia-materials-studio/ [Accessed: 15 June 2020]

[23] Putz H, Brandenburg K. Match!—Phase Analysis Using Powder Diffraction, Version 3.x, Crystal Impact. Bonn, Germany, GbR, Kreuzherrenstr. **102**:53227. Available from: https://www.crystalimpact.de/match

[24] Hildebrand JH. Solubility. Journal of the American Chemical Society. 1916; **38**:1452

[25] Greenhalgh DJ, Williams AC, Timmins P, York P. Solubility parameters as predictors of miscibility in solid dispersions. Journal of Pharmaceutical Sciences. 1999;**88**(11): 1182-1190. DOI: 10.1021/js9900856

[26] Forster A, Hempenstall J, Tucker I, Rades T. Selection of excipients for melt extrusion with two poorly water-soluble drugs by solubility parameter calculation and thermal analysis. International Journal of Pharmaceutics. 2001; **226**(1–2):147-161. DOI: 10.1016/S0378-5173(01)00801-8

[27] Li B et al. Molecular dynamics simulation of CO_2 dissolution in heavy oil resin-asphaltene. Journal of CO_2 Utilization. 2019;**33**:303-310

Chapter 6

Rare Earth Elements in New Advanced Engineering Applications

Monika Duchna and Iwona Cieślik

Abstract

From an engineering approach, rare earth elements (REE) have the extra potential to modify modern engineering in an extraordinary way. Their peculiar optical, mechanical, electronic, and magnetic properties have been used for years and even open up wider possibilities for using rare earth elements. With advances in all fields of engineering, it is predictable that the rare earth elements will play a crucial role. The use of the rare earth elements permits many new advances, including digital and magnetic technologies operating at reduced energy consumption, higher efficiency, miniaturization, speed, and durability. The REEs are particularly crucial components in clean energy applications, which is especially important in the fight against global warming. For these reasons, the rare earth elements will become essential components in the technological revolution in the second quarter of the twenty-first century.

Keywords: REE engineering, permanent magnets, clean energy, phosphors, bioimaging

1. Introduction

Rare earth elements (REE) are extremely important components in high technology, which is why they are sometimes referred to as "vitamins" of the modern economy or industry [1, 2]. With the development of high technology, the demand for rare earth elements increases year on year. It is especially visible in the case of searching for clean energy sources or in the development of various types of electronic devices. Rare earth elements are important components in advanced technologies such as smartphones, computers, TVs, LEDs, hard drives, or elements used for the production of clean energy, such as magnets in wind turbines. Therefore, an increase in REE production is observed worldwide (**Figure 1**). In 2021, the extraction of rare earth elements in the world reached 280,000 tons, which was a significant increase compared to 2018, when 110,000 tons less were extracted. China is the largest producer of REE [3]. In 2021, they have produced 168,000 tones of these elements, while the United States came second in this ranking with a much lower production of 43,000 tones [4].

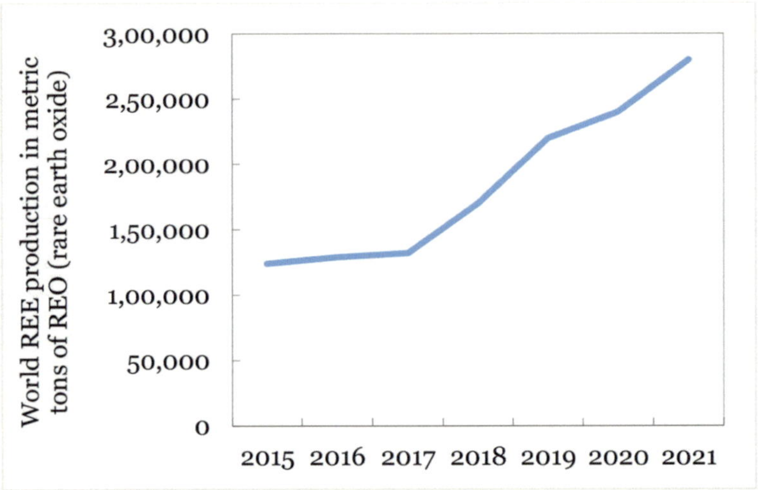

Figure 1.
World mine production of REE (after Ref. [3]).

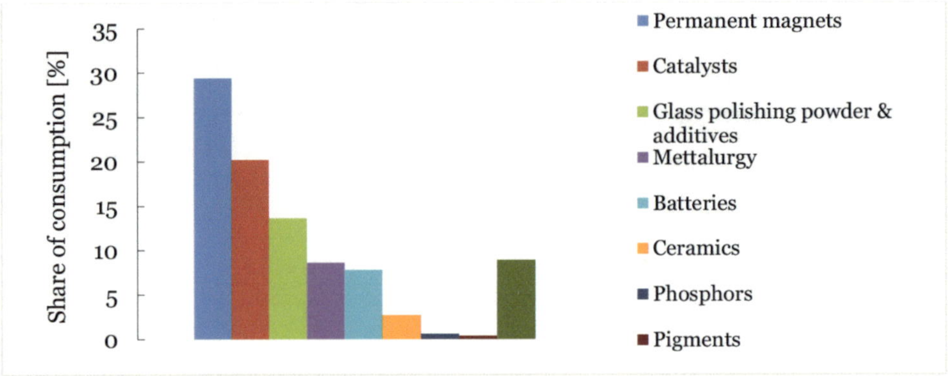

Figure 2.
Distribution of REE consumption worldwide in 2020, by end-use applications (after Ref. [5, 6]).

In the natural environment, REEs do not occur as single native metals, such as gold or silver, they occur together in many ores or minerals as secondary or major constituents [1]. The most economically viable methods of extracting these elements from the ores, as well as separating them from the rest of the components, are being sought. The separation of REEs is difficult and costly due to the similar chemical properties of these elements. In addition to searching for new, more economical methods of their separation, the issue of their reuse - recycling is also very important. Each of the rare earth elements has unique chemical, optical, mechanical, electronic, or magnetic properties and can therefore be used in advanced engineering applications such as permanent magnets, luminescent materials, metallurgy, batteries, catalysts, ceramics, pigments, phosphors, nuclear industry, medicine, and nanotechnology.

The growing importance of these elements in the development of new technologies has made them sought-after raw materials worldwide. It is very important to maintain a proper supply chain for highly developed economies that want to participate in the development of highly advanced technologies. The distribution of the use of these elements in the world, in particular, advanced applications in 2020 is presented in **Figure 2**.

2. Applications of REE

Rare earth elements are applied in many different fields (**Figure 2**). They are used in mature industries such as catalysts, glass production, and metallurgy. However, they play an increasingly important role in the applications of high technology, which include batteries, ceramics, and permanent magnets. In the case of mature markets, elements such as lanthanum and cerium are most often used, in contrast to high-technologies, where dysprosium, neodymium, and praseodymium are used [7]. Permanent magnets production is the largest and most important end use of REE, accounting for over 29% of the total consumption of these elements in 2020 [5]. The main uses of rare earth elements are presented in **Table 1**. Above mentioned applications, there are still areas that use rare earth elements in new ways, such as nanotechnology, which can be employed e.g. in medicine. This chapter focuses primarily on the advanced applications of these elements in nanotechnologies as well as in the most important application in the world today, which are permanent magnets.

REE	Symbol	Applications
Scandium	Sc	High-strength Al-Sc alloys, electron beam tubes
Yttrium	Y	Phosphors for fluorescent lighting and liquid crystal displays (LCDs), capacitors, radars, lasers, superconductors, glasses
Lanthanum	La	Battery alloys, phosphors, glasses, ceramics, car catalysts, lasers, pigments, accumulators
Cerium	Ce	Catalysts, phosphors, ceramics, glasses, pigments
Praseodymium	Pr	Permanent magnets, photographic filters, ceramics, glasses, pigments
Neodymium	Nd	Permanent magnets, catalysts, lasers, pigment for glass and ceramic
Promethium	Pm	Miniature nuclear batteries, phosphors
Samarium	Sm	Permanent magnets, reactor control rods
Europium	Eu	Fluorescent lighting and LCDs
Terbium	Tb	Lighting and displays phosphors, permanent magnets
Dysprosium	Dy	Permanent magnets, lasers, lighting, nuclear industry
Holmium	Ho	Magnets, lasers, nuclear industry
Erbium	Er	Lasers, optical fibers, glass colorant, the nuclear industry
Ytterbium	Yb	Solar panels, fiber optics, lasers, metallurgy, nuclear medicine
Lutetium	Lu	X-ray phosphors
Thulium	Tm	Magnets, electron beam tubes
Gadolinium	Gd	Nuclear fuel bundles, medical imaging, electronics

Table 1.
Applications of REEs (after Refs. [7–10]).

3. Permanent magnets

Rare earth metals are key elements of permanent magnets, which are used as components in clean energy applications such as wind turbines or motors in electric vehicles. Permanent magnets convert electrical energy into mechanical energy (motors) or inversely (generators) by generating a magnetic field [11, 12]. In recent years, the development of wind energy technology has been observed, which is related to the desire to increase the percentage share of renewable energy sources in energy production in the world and, thus reduce carbon dioxide emissions. **Figure 3** shows the periodic table of elements with highlighted rare earth metals used for permanent magnets. As can be seen from the figure, among rare earth metals, lanthanides are used to produce permanent magnets. These elements are characterized by good magnetic properties. The properties of lanthanides are greatly influenced by the *f* orbital, which is also a factor that distinguishes them from transition metals. This orbital is located in the atomic core and therefore does not participate in chemical bonding. Since there are 7 orbitals of the 4*f* type, the number of unpaired electrons can be as high as 7. This results in the formation of large magnetic moments in lanthanide compounds [14]. Lanthanides are paramagnetic except for La^{3+} ($4f^0 5d^0 6s^0$), Ce^{4+} ($4f^0 5d^0 6s^0$), Lu^{3+} ($4f^{14} 5d^0 6s^0$) and Yb^{2+} ($4f^{14} 5d^0 6s^0$), they are diamagnetic – they have no unpaired electrons [15]. Examples of electron configurations of paramagnetic lanthanides are presented in **Figure 4**.

Magnets containing rare earth metals were discovered in the 1960s in the United States. They were based on samarium and the transition metal cobalt (SmCo). The first type of this kind of magnet was SmCo5 which had properties suitable for permanent magnets including large uniaxial magnetocrystalline anisotropy, comparatively high saturation magnetization, high Curie temperature, and maximum energy

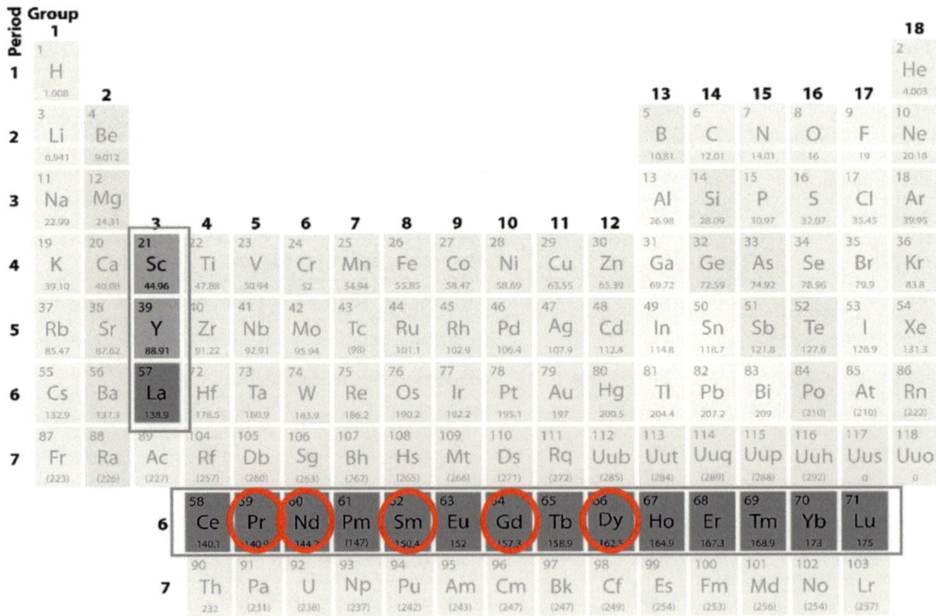

Figure 3.
Location of rare earth metals on the periodic table, elements used in permanent magnets are marked in red [13].

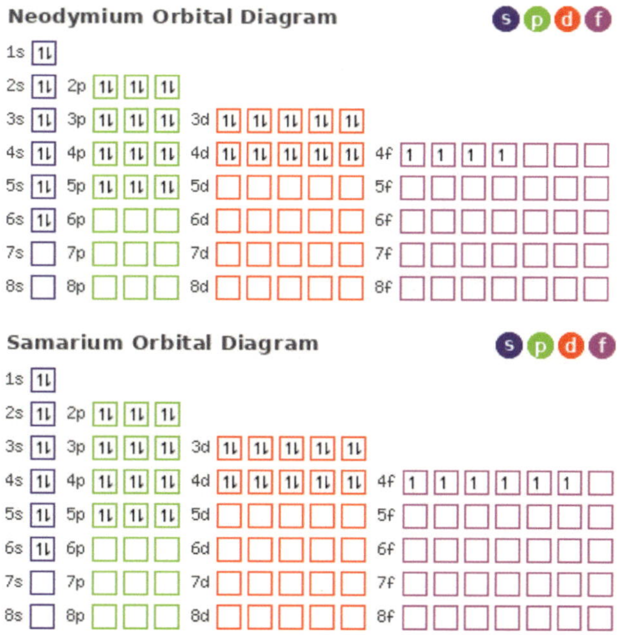

Figure 4.
Electron configuration on orbital diagrams for rare earth elements neodymium and samarium [16].

product [$(BH)_{max}$] with the value of 160 kJ/m^3 [17]. Next, to obtain higher magnetizations, a compound with a stoichiometry of Sm_2Co_{17} was designed. Nevertheless, despite obtaining a higher saturation magnetization and Curie temperature compared to SmCo5, the field anisotropy in the case of Sm_5Co_{17} was smaller [17]. The major disadvantage of these magnets was the relatively high cost of Sm and Co, which resulted that in the 1980s, researchers in Japan and the United States designed a new type of magnet containing the rare earth metal neodymium as well as iron and boron ($Nd_2Fe_{14}B$) [2, 17]. It was cheaper and at the same time stronger compared to the SmCo magnet. The development of permanent magnets in the world is shown in **Figure 5**.

Rare earth magnets are distinguished by excellent magnetic properties determined by high induction and coercive force. They are much more powerful per unit mass and volume than other types of magnets [19]. This is related primarily to the high magnetocrystalline anisotropy. The resistance of the crystal lattice to a change in the direction of magnetization gives these materials a very high magnetic coercivity (resistance to demagnetization), which may be attributed to a strong demagnetizing field in the finished magnet than does not reduce the magnetization of the material [2].

The development of permanent magnet technology has been achieved through the use of alloys containing rare earth elements such as Nd, Sm, Gd, and Pr. The introduction of additives in the form of these elements made it possible to produce magnets characterized by a lower mass, smaller dimensions, and high strength at the same time [20]. As a consequence, it made it possible to significantly reduce the size of various types of electronic devices and their components. Compared to other permanent magnets, rare earth magnets can significantly reduce the size and weight of generators used for the production of clean energy in wind turbines. Such magnets can increase efficiency above 20%, which is also very important from an economic point of view [8].

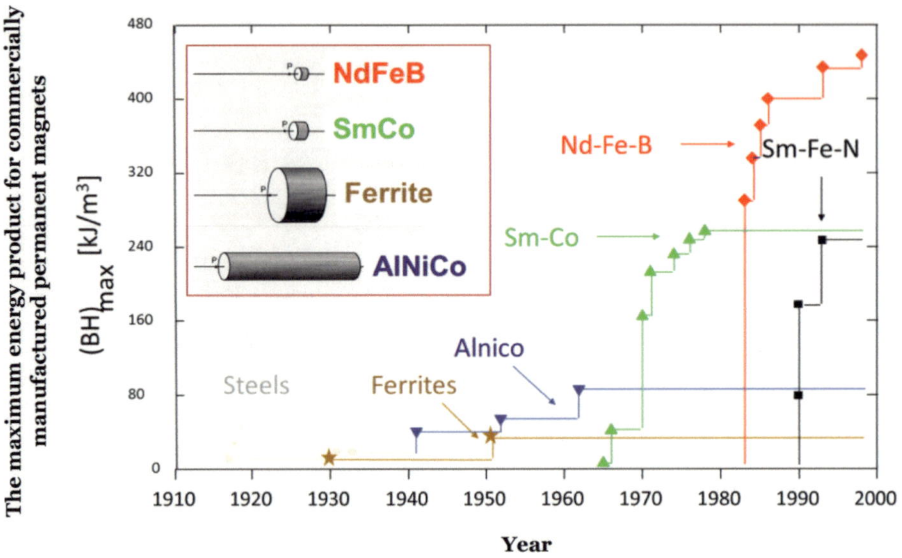

Figure 5.
Development of permanent magnets [18].

Rare earth magnets are employed in applications that require a high magnetic field in difficult operating conditions such as high temperature and high demagnetization forces. In SmCo magnets, samarium is the dominant rare earth metal and cobalt is the primary transition metal, often used with iron, zircon, and copper. The quantity of rare earth elements in such magnets ranges from 25 to 35% by weight. Samarium cobalt magnets are used at high temperatures [2]. However, some disadvantages, include their brittleness which limited the size of the magnet and thus seriously restricts the possibility of their use in certain applications, e.g. in car motors [8]. Another rare earth element, gadolinium, is used to achieve a near-zero change in the residual induction over a wide temperature range. The difference between the SmCo and GdCo magnets is that with increasing temperature, the residual induction decreases in the first case and increases in the second, although from a significantly lower initial value. The combination of SmCo and GdCo allows to obtain magnets with high stability. Magnets based on samarium and cobalt have been used, among others e.g. in generators, actuators, or medical devices [2].

NdFeB magnets (also known as Neo or NIB magnets) [21] are now largely replacing SmCo magnets [22]. NdFeB magnets are more powerful than those of SmCo, and offer the strongest magnetic flux density, therefore are widely used in clean energy technologies such as wind turbines and electric vehicle motors [23]. The neodymium content in these materials is about one-third. The typical chemical composition of NIB magnets is 26.7% Nd, 72.3% Fe and 1.0% B. Although the content of Nd and Fe elements may differ somewhat in commercially used magnets [21]. Frequently other rare earth element praseodymium (Pr) replaced neodymium in these compounds [2, 24]. This is related, among others to the reduction of their production costs.

Previously mentioned use of such magnets are wind turbine generators. Wind turbines require the use of generators with greater power, therefore magnets such as NdFeB are widely used in this type of application. Thus it is possible to substitute the mechanical gears in wind turbines with permanent magnet generators. The advantages of using permanent magnets in wind turbine generators are a reduction in the total weight of the turbine as well as a reduction in the number of moving parts, which results in an increase in the efficiency and reliability of these turbines. The benefits of using permanent magnet generators are especially important for offshore installations, where reliability is a priority. Which is associated with the high cost of repair and maintenance of such turbines [8]. The magnet in a large wind turbine contains as much as 260 kg (or more) of neodymium [25]. Due to their properties, these magnets have also found application in other technologies used to produce renewable energy under the ocean floor and by waves [8].

In addition to wind turbines, where magnets play a key role, they also dominate the market for the production of motors for electric cars. In this case, the size and weight of these components are also important, and they must be adapted to the previously designed engine parameters [8, 22]. As a consequence of the high magnetic strength of NdFeB magnets, they can produce a lot of energy in relation to their weight and size. This makes them ideal for applications that need a high energy-to-weight ratio, such as electric vehicle motors.

It is possible to introduce small amounts of additional ingredients in the form of heavy rare earth metals or other metals, which may improve the properties of the Neo magnets. A very important addition is dysprosium, which increases intrinsic coercivity and resistance to demagnetization of these magnets and in consequence enables the use of them at higher temperatures [22, 26]. The addition of dysprosium is usually from 2 to 5% [2, 8]. As mentioned before NdFeB magnets are requested for applications of small or large motors and generators. Small motors are used e.g. in power disc drivers in computers, while large in electric vehicles. In electric car motors, up to 200 g of neodymium and 30 g of dysprosium are used whereas wind turbine generators can contain 1 ton of neodymium per megawatt of electrical power generated [23, 27].

Another common application of this type of magnets is electronics as well as in lasers and telecommunications, there are 12 times stronger than standard iron magnets [28].

4. Using optical properties in new technologies

The optical properties of lanthanides are still number one in rare earth (RE) ions applications since the late nineteenth century. In 1964 two scientists discovered possibilities for this elements group so-called red phosphors. The red phosphors were used mainly in the TV screens production process. That technology used a mixture of two metal elements from the sub-shell "*d*" group of the periodic table as europium and yttrium ions. As a result, it became possible to obtain a TV color [29]. These excellent optical properties of RE are still used in several of new technologies such as modern lighting displays, photodynamic therapy, and biodetection. These technologies give an opportunity to develop numerous industry sectors.

The reason of the great array optical properties is an unusual electronic structure on the sub-shell "*d.*" This characteristic electronic structure is related to the existence of an important phenomenon called luminescence. That phenomena depend on three mechanisms of electron transfer: 4f-4f or 5d-4f and charge transfer (CT).

Moreover in the CT mechanism, the easily oxidizing or reducing ions like Ce^{3+}, Pr^{3+}, Tb^{3+}, and Eu, Tm, Yb entail depending the transitions intensity broad bands produces from the symmetry of the surroundings of the RE ion.

RE ions due to base on their 4f-4f or 5d-4f energy transfer and thus the emission of powerful light sources in a wide range [29]. The 5d-4f energy transfer largely depends on the ligands surrounding the RE ions. Hence it becomes more reactive unluckily. However, the 4f-4f transitions are the most commonly used in the range of visible light the lifetime of the luminescence range from tens of microseconds to single milliseconds. In addition, there is the shielding effect of the filled $5s^2$ and $5p^6$ shells which contributes to the environment protection. Thus external factors have no influence on them. This makes them much better candidates suited to applications in the field of optics than organic compounds.

In the last years, a large development of nanotechnology in optics is observed. Among the interesting new materials, there are also nanomaterials containing RE ions. Those modern nanomaterials are called witching others: fluorescent nanoparticles, quantum dots, or nanocolloidal metallic nanoparticles. The nano-luminophores and nanoplatforms demonstrated desirable properties and were devised functionally in the last years [30]. The most commonly used RE elements in nanotechnology are: Eu^{3+}, Tb^{3+}, and co-doped phosphors Yb^{3+}-Tm^{3+}, Yb^{3+}-Er^{3+}, or Yb^{3+}-Ho^{3+}.The emission of RE ions covers the entire range from UV (Gd^{3+}) by visible light (Tb^{3+}, Tm^{3+}, Sm^{3+},) to the NIR range (Yb^{3+}, Er^{3+}, Pr^{3+} Nd^{3+}). However, two RE ions: La^{3+} and Lu^{3+} are unable to give 4f-4f transfer and do not emit light. A lifetime of emission light depends on the size of the crystallites. It was noted that the light emission in phosphors is growing with the reduction increasing of the phosphors crystalline [31]. This could apply e.g. nanomarkers doped with RE. These nanomaterials can actively influence on the cells. The environment of cells can be heating locally and formation of free-radicals at once. That phenomena have a significant role in photodynamic therapy (PDT) and hyperthermia. The most important is using NIR range light to excite the phosphors which allows to less invasive and deep penetrations of cancer cells.

Currently, however, trials are underway to extend the therapeutic indications for extracorporeal analysis as screening test genetic diseases, detection of infections caused by different microorganisms, and the presence of impurities of bacteria in water and food products [32]. Therefore, nanophosphors becoming increasingly fashionable in various ranges of applications. Their unique physical and chemical properties allow for quick and easy applications in a new range of applications notably in medicine range.

Nevertheless, one of the most developed directions is biological applications such as biodetection and bioimaging [33]. The extraordinary spectroscopic properties of RE as stable luminescence, up-conversion ability, and a narrow band of light emission resulted in an increased interest in them. Recent decades have seen the development of research and an increased interest in RE in nanoform in various biological applications. A pie chart **Figure 6** shows all characteristic biological application RE ions which are the most popular in the last years.

Among the widespread type of biodetection and bioimaging are phosphorus markers. These makers give opportunities for a wide range of applications in medicine as drags transfers, photodynamic therapy, in vivo imaging of biochemical reactions in real-time, cancer therapy, or imaging DNA mutations. From a medical viewpoint, these opportunities give wide and new perspectives.

The use of RE in multiplexing also plays an important role. The multiplexing with RE being capable to for the analysis of multiple analytes in one sample. This possibility allows to apply RE in genetic tests or new studies on cancer drugs.

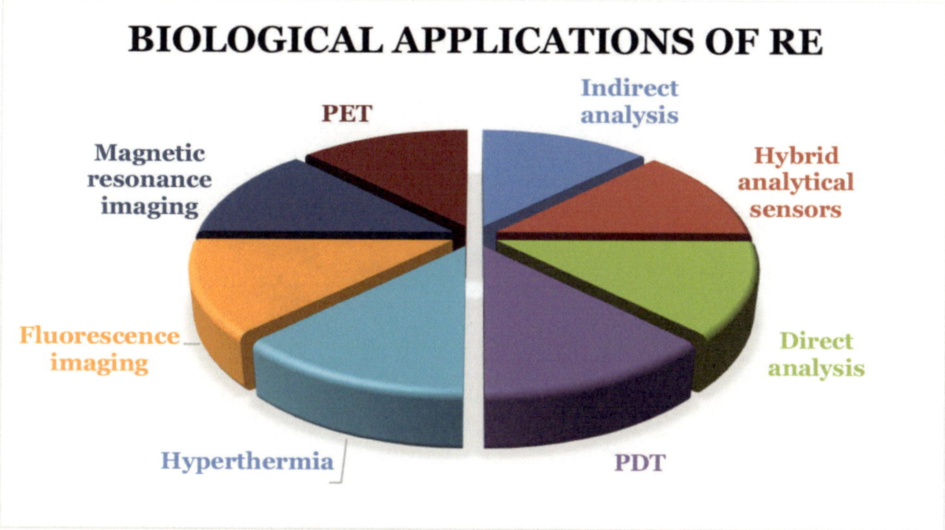

Figure 6.
*New way of biological applications of RE elements. *PET, positron emission tomography; *PDT, photodynamic therapy.*

Interesting results are silica-based or polymeric-based hybrid phosphors. The silica in SiO form and polymeric in PEG form plays a role of ligands. The matrix where RE ions are embedded can be various. The type of matrix usually used in phosphors for biological applications were shown in **Table 2**. As can be observed, all of these list medical applications are based on phosphors or nanophosphors.

Diagnostics and photodynamic therapy in the use of lanthanides are one of the most modern methods of imaging and cancer therapy of PDT. The PDT method is based on the up-convergence mechanism. This process generally depends on the absorption of successive photons (**Figure 7**). For example, the β-NaYF4 matrix doped with Er^{3+} and Yb^{3+} is characterized by about 10^5 times greater efficiency than organic compounds, e.g. rhodamine 6G [39].

Matrics	RE ions	Ligands	Applications
$NaYF_4$	Yb-Tm, Yb-Ho Yb-Ho-Ce	—	In vivo imaging
$NaYF_4$	Yb^{3+}, Er^{3+}	$NaYF_4$/SiO2	Photodynamic therapy
Y_2O_3	Er^{3+}	PEG	Cancer cells detection
$NaYF_4$	Nd^{3+}	—	Hiperthermic
LaF_3	Ce^{3+}, Tb^{3+}	PSS, PAH	Glucose marking
$NaYF_4$	Yb^{3+}, Er^{3+}, Gd^{3+},	^{18}F, citric / oleic acid	Magnetic resonance Emission tomography
Gd_2O_3, Fe_3O_4	Eu^{3+}	DNA	Detection of DNA mutations
LaF_3	Yb^{3+}, Er^{3+}	SiO	Transportation of drugs

Table 2.
A selected list of the luminescence materials with the most popular RE ions used in biological experiments [34–38].

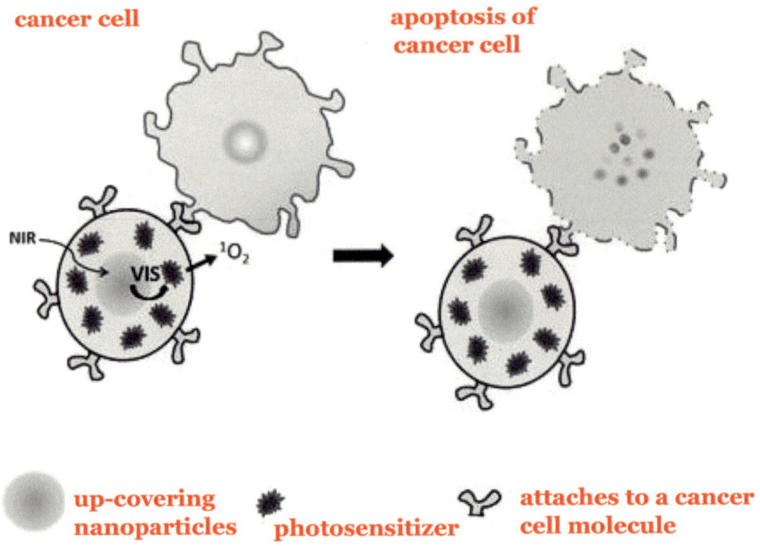

Figure 7.
Scheme of PDT mechanism based on nanoluminophores doped with RE–the matrix with nanoparticles doped RE ions is excited by infrared light NIR and the core is emitting visible light VIS. The VIS light exciting photosensitizers molecules and generate reactive oxygen form $^1O_2$. The cancer cells undergo apoptosis [30].

On the other hand, biological imaging is one of the most important applications of RE elements in a science experiments, various electronic devices, and medicine. NaYF4:Er^{3+}, Tm^{3+}, Ho^{3+} nanocrystallite was used for cancer cells imaging [40]. The up-conversion mechanism of RE ions in special matrix was also used. The deep penetration excitation infrared light of cells and absence of photobleaching problem of nanoparticles especially. In this case doping of several RE ions of matrix NaYbF4 gives the opportunity to obtain wavelength ranges with different colors.

However, biological or medical applications are not the only ones where RE phosphors are used. The up-conversion is using also to various sensors production with nanoluminophores doped with RE ions as pH sensors, oxygen sensors, ammonia sensors, or carbon dioxide sensors [41–43]. The most frequently the ytterbium, thulium, or europium ions are used for this purpose. The first designed oxygen sensors NIR light excited was NaYF4:Yb^{3+}, Tm^{3+}.

Silicate phosphor KBaScSi2O7:Ce^{3+} (KBSS:Ce^{3+}) can emit cyan light with an emission peak at 509 nm under n-UV light excitation (300–400 nm). Those optical properties indicate that KBSS:Ce^{3+} phosphor is favorable in LED and field emission display applications [44].

5. Conclusions

Due to the unique magnetic, luminescent, chemical, and physical properties of rare earth elements, are essential ingredients for many high-technology applications, and there still will be observed demand for them in the future. Especially since their significant use is related to the production of clean energy, which nowadays is a very important issue. The development of wind energy will continue to drive demand for REE used in wind turbine generators. Converting combustion engine cars w to electric vehicles will also increase the need for magnets and rare earth batteries. Of course, the search is ongoing and materials that could replace or reduce the amount of REE in individual applications are considered. However, due to their superior properties, we are currently unable to replace them in some usage because it would involve a deterioration of the quality and efficiency of the components in which they are used. There are still applications where the potential of rare earth elements has not been fully exploited and we will certainly observe their further development.

Author details

Monika Duchna[1,2]* and Iwona Cieślik[1]

1 National Centre for Nuclear Research, Otwock, Poland

2 Faculty of Materials Science and Engineering, Warsaw University of Technology, Warsaw, Poland

*Address all correspondence to: monika.duchna@ncbj.gov.pl

References

[1] Balaram V. Rare earth elements: A review of applications, occurrence, exploration, analysis, recycling, and environmental impact. Geoscience Frontiers. 2019;**10**:1285e1303

[2] Dent PC. Rare earth elements and permanent magnets (invited). Journal of Applied Physics. 2012;**111**:07A721

[3] U.S. Geological Survey. Mineral Commodity Summaries, Rare Earths. 2015-2022

[4] Pistilli M. Top 10 Countries for Rare Earth Metal Production. Updated 2022. Available from: https://investingnews.com/daily/resource-investing/critical-metals-investing/rare-earth-investing/rare-earth-metal-production/

[5] Garside M, editor. Distribution of Rare Earth Element Consumption Worldwide in 2020, By End Use; 2022. Available from: https://www.statista.com/statistics/604190/distribution-of-rare-earth-element-consumption-worldwide-by-end-use/

[6] Government of Canada. Rare Earth Elements Facts. 2022. Available from: https://www.nrcan.gc.ca/our-natural-resources/minerals-mining/minerals-metals-facts/rare-earth-elements-facts/20522

[7] Charalampides G, Vatalis KI, Apostoplos B, Ploutarch-Nikolas B. Rare earth elements: Industrial applications and economic dependency of Europe. Procedia Economics and Finance. 2015;**24**:126-135

[8] Smith Stegen K. Heavy rare earths, permanent magnets, and renewable energies: An imminent crisis. Energy Policy. 2015;**79**:1-8

[9] US DOE (US Department of Energy). Critical Materials Strategy. 2011. Available from: http://energy.gov/sites/prod/files/DOE_CMS2011_FINAL_Full.pdf

[10] US EPA (US Environmental Protection Agency). Rare Earth Elements: A Review of Production, Processing, Recycling, and Associated Environmental Issues. 2012. Available from: http://nepis.epa.gov/Adobe/PDF/P100EUBC.pdf

[11] Zijlstra H. Application of permanent magnets in electromechanical power converters; the impact of Nd-Fe-B magnets. Journal de Physique, Colloque. 1985;**46**:C6-3-C6-8

[12] Nakamura H. The current and future status of rare earth permanent magnets. Scripta Materialia. 2018;**154**:273-276

[13] Rare Element Resources Ltd; 2016. Available from: https://www.rareelementresources.com/rare-earth-elements#.YxB8MNPP2Uk

[14] Starowicz M. AGH University of Science and Technology in Krakow. Rare Earth Metals, Lanthanides. 2022. e-Book Available from: https://pre-epodreczniki.open.agh.edu.pl/tiki-index.php?page=Lantanowce#lantanowce

[15] Venkatraman G, Verma UN. Stereochemistry, Nuclear Chemistry, Lanthanides and Actinides Group Theory and its Applications. Unit - 3 Lanthanides and Actinides; 2009. Available from: http://www.nou.ac.in/econtent/Msc%20chemistry%20paper%202/MSc%20Chemistry%20Paper-II%20Unit-3.pdf

[16] Neodymium and Samarium Orbital Diagrams. Available from: https://www.chemicalaid.com/

[17] Sugimoto S. Current status and recent topics of rare-earth permanent magnets. Journal of Physics D: Applied Physics. 2011;**44**:064001

[18] Gutfleisch O, Willard MA, Brück E, Chen CH, Sankar SG, Ping Liu J. Magnetic materials and devices for the 21st century: Stronger, lighter, and more energy efficient. Advanced Materials. 2011;**23**:821-842. DOI: 10.1002/adma.201002180

[19] Van Gosen BS, Verplanck PL, Long KR, Gambogi J, Seal II RR. The Rare-Earth Elements: Vital to Modern Technologies and Lifestyles, U.S. Geological Survey. 2014. Fact Sheet 2014-3078

[20] Haxel GB, Hedrick JB, Orris GJ. Rare Earth Elements—Critical Resources for High Technology, U.S. Geological Survey. 2002. Fact Sheet 087-02

[21] Tripathy PK, Mondal K, Khanolkar AR. One-step manufacturing process for neodymium-iron (magnet-grade) master alloy. Materials Science for Energy Technologies. 2021;**4**:249-255

[22] Zhou B, Li Z, Zhao Y, Zhang C, Wei Y. Rare earth elements supply vs. clean energy technologies: New problems to be solve. Gospodarka surowcami mineralnymi – Mineral Resources Management. 2016;**32**:29-44

[23] Goonan TG. Rare Earth Elements—End Use and Recyclability, U.S. Geological Survey. 2011. Scientific Investigations Report 2011-5094

[24] Hui M. Why Rare Earth Permanent Magnets are Vital to the Global Climate Economy. 2021. Available from: https://qz.com/1999894/why-rare-earth-magnets-are-vital-to-the-global-climate-economy/

[25] Burchard-Dziubińska M. Strategiczna rola metali ziem rzadkich w gospodarce opartej na wiedzy (Strategic role of rare-earth metals in the knowledge-based economy. Gospodarka w Praktyce i Teorii. 2014;**1**(34):21. Available from: https://dspace.uni.lodz.pl/xmlui/bitstream/handle/11089/11867/02%20Burchard-Dziubinska.pdf?sequence=1&isAllowed=y

[26] BGS. Rare Earth Elements. British Geological: Survey; 2010

[27] Keane E. Neodymium Magnets Provide Key to Understanding Rare Earth Trends: Seeking Alpha. 2009. Available from: https://seekingalpha.com/instablog/345817-eamon-keane/9675neodymium-magnets-provide-key-to-understanding-rare-earth-trends

[28] Daigle B, DeCarlo S. Rare Earths and the U.S. Electronics Sector: Supply Chain Developments and Trends. 2021. U.S. International Trade Commission (USITC), No. ID-21-075

[29] Li LZ, Yan B, Lin L-X, Zhao Y. Solid state synthesis, microstructure and photoluminescence of Eu^{3+} and Tb^{3+} activated strontium tungstate. Journal of Materials Science: Materials in Electronics. 2011;**22**:1040-1045

[30] Misiak M, Prorok K, Bednarkiewicz A. Biological application of lanthanide doped nanomaterials. Wiadomości chemiczne. 2012;**66**:5-6

[31] Bednarkiewicz A, Mączka M, Strek W, Hanuza J, Karbownik M. Size dependence on infrared spectra of NaGdF4 nanocrystals. Chemical Physics Letters. 2006;**418**:75

[32] Corstjens PL, Li S, Zuiderwijk M, Kardos K, Abrams WR, Niedbala RS, et al. Infrared up-converting phosphors

for bioassays. IEEE Proceedings of Nanobiotechnology. 2005;**152**:64

[33] Cieślik I, Bolek T, Woźniak MJ, Majchrowski A, Hirano S, Budzianowski A. Potassium gadolinium tungstate nanocrystals doped with holmium ions as candidates for optical imaging. Applied Surface Science. 2018;**446**:139-144

[34] Yang Y, Qu Y, Zhao J, Zeng Q, Ran Y, Zhanng Q, et al. Fabrication of and drug delivery by an upconversion emission nanocomposite with monodisperse LaF3:Yb,Ercore/mesoporous silica shell structure. European Journal of Inorganic Chemistry. 2010;**2010**:5195

[35] Qian LI, Zhou LH, Too HP, Chow GM. Gold decorated NaYF4:Yb,Er/NaYF4/silica (core/shell/shell) upconversion nanoparticles for photothermal destruction of BE(2)-C neuroblastoma cells. Journal of Nanoparticle Research. 2010;**13**:499

[36] Zhou J, Sun Y, Du X, Xiong L, Hu H, Li F. Dual-modality in vivo imaging using rare-earth nanocrystals with near-infrared to near-infrared (NIR-to-NIR) upconversion luminescence and magnetic resonance properties. Biomaterials. 2010;**31**:3287

[37] Zako T, Nagata H, Terada N, Utsumi A, Sakono M, Yohda M, et al. Cyclic RGD peptide-labeled upconversion nanophosphors for tumor cell-targeted imaging. Biochemical and Biophysical Research Communications. 2009;**381**:54

[38] Zhou J, Yu M, Sun Y, Zhang X, Zhu X, Wu Z, et al. Fluorine-18-labeled Gd3+/Yb3+/Er3+ co-doped NaYF4 nanophosphors for multimodality PET/MR/UCL imaging. Biomaterials. 2011;**32**:1148

[39] Park YI, Kim JH, Lee KT, Jeon KS, Na HB, Yu JH, et al. Nonblinking and nonbleaching upconverting nanoparticles as an optical imaging nanoprobe and T1 magnetic resonance imaging contrast agent. Advanced Materials. 2009;**21**:4467

[40] Wang M, Mi C, Zhang Y, Liu J, Li F, Mao C, et al. NIR-Responsive Silica-Coated NaYbF4:Er/Tm/Ho Upconversion Fluorescent Nanoparticles with Tunable Emission Colors and Their Applications in Immunolabeling and Fluorescent Imaging of Cancer Cells. Journal of Physical Chemistry C. 2009;**113**:19021

[41] Sun LN, Peng H, Stich MI, Achatz D, Wolfbeis OS, pH sensor based on upconverting luminescent lanthanide nanorods. Chem Commun (Camb). 2009;**33**:5000-5002

[42] Achtz DE, Meier RJ, Fischer LH, Wolfbeis OS. Luminescent sensing of oxygen using a quenchable probe and upconverting nanoparticles. Angewandte Chemie, International Edition. 2011;**50**:260

[43] Mader HS, Wolbies OS. Optical ammonia sensor based on upconverting luminescent nanoparticles. Analytical Chemistry. 2010;**82**:5002

[44] Zhou Y, Chen H, Wang Y. Cyan broad-band emission phosphor with scandium silicon multiple-ring structure for white light-emitting diodes and field emission display. Inorganic Chemistry. 2021;**60**(12):8870-8879

Chapter 7

An Overview of Radon Emanation Measurement System for South African Communities

Moses Radebe and Manny Mathuthu

Abstract

The aim of the study is to evaluate radon emanation levels in South African communities and to implement possible strategies to reduce radon levels in order to minimize potential health hazards. The major contributing factor to high levels of radon is the history of mining. To precisely measure emanation levels of radon indoors and outdoors, calibrated radon detectors are necessary. In this study, areas of high radon emanation levels are spotted, and based on the radon emission point or entry points in buildings, applicable and possible mitigation strategies are discussed for implementation.

Keywords: radon emanation, mitigation strategies, calibration techniques, radon chamber, NORM

1. Introduction

Radon is the most contributing source of ionizing radiation to human beings, in the atmosphere. It is a colorless, invisible, undetectable gas to human senses that causes lung cancer especially in areas where it is abundant. Areas of high levels of radon include underground uranium mines and locations close to mine tailings of which its buildings are non-ventilated. Lubin et al. [1] state that about 40% of deaths in mine workers are linked to the radon emanation in underground mines, in United States, radon is the second most contributing source to deaths due to lung cancer [2]. The International Commission on Radiological Protection recommends that the radon concentration level in dwellings should not exceed 300 Bq/m^3 [3]. In South Africa, radon from areas of possible high radon emanation is a concern due to the history of mining, disposal of uranium tailing, use of mine waste for building materials [4]. Areas of high radon emanation include:

- Large Tailing Storage Facilities in Gauteng Province. Mine Tailings mismanagement run off to the public.

- Granite hill in Saldanha Bay area and houses built on granite bedrock in Paarl, Western Cape.

- Houses built by uranium ore stockpiles in Karoo, Western Cape.

Mitigation strategies to reduce radon to acceptable levels to the public and mine workers are of necessity for implementation. Strategies such as reducing radon levels in underground mines by ventilation systems, reducing radon emanation in tailing dams by pouring 2 m of clay and topsoil to trap radon, and identifying radon entry points for sealing in houses or mines. To evaluate the effectiveness of the mitigation system to reduce radon to ensure safe public health, a properly calibrated radon detection device is necessary to ensure accurate exposure levels of radon in workplaces and houses [5]. With a calibrated radon detector, results of measurement before and after application of a mitigation system will absolutely indicate the rate of reduction of radon emanation. Therefore, buildings can be fixed if high radon emanations are detected, shortening the statistics of lung cancer deaths due to excessive exposure to radon.

The most common method of calibrating the radon detectors is exposing them to a steady flow of radon concentration from a standard radon source in an airtight radon chamber, under controlled environmental parameters [6]. A facility to calibrate radon detectors in a designed radon chamber at the Centre of Applied Radiation Science and Technology is in development phase.

In the following subsections, radon radiological properties, its transport behavior in the environment, and mitigation system will be introduced.

1.1 Radon radiological properties

^{222}Rn is a noble radioactive isotope of atomic number 86, and it originates from the decay chain of ^{238} U as one of its decay product, ^{226}Ra transforms into ^{222}Rn by alpha particle emission as seen in **Figure 1**.

Radon alpha particles travel a distance of 4–7 cm with an energy of 5.5 MeV. Alpha particles' energy from radon daughter nuclei is more than that of radon, the parent. Polonium (^{218}Po) with half-life of 3.05 minutes has energy of 6.0 MeV, while polonium (^{214}Po) has 7.7 MeV. Hence, radon daughter nuclei are dangerous and pose health risk in an enclosed space such as offices and houses.

1.2 Transport of radon in the environment

Radon is found in rocks and soils, and its abundance in the environments depends on moisture, porosity, and the activity of uranium and radium in soils and rocks. The concentration of radon in soil differs due to the concentration of its parent radionuclide and the ability of the soil to emanate radon.

Factors that affect the mobility of radon in the soil include porosity, moisture content of soil, and permeability. Hosoda [8] found that the moisture content of soil in the range of 0–8% in a rectangular volume of 2840 cm^3 increases radon emanation but, moisture content that is above 8% decreases radon emanation. Permeability of soil such as gravel and sand allows the transport of radon gas from several depths in the ground as seen in **Figure 2**. Impermeable soil such as clay and silt has low porosity, and therefore, the transport of radon is small. Decreased levels of radon in clays are due to the amount of water content in them, but in the dry clays with cracks, radon migration is more in non-cracked clay [10, 11].

Rocks with varying degrees of radium activity content are the source of radon in groundwater. The measure of ^{222}Rn activity differs from surface water of lakes or rivers, as radon in underground water is not mobile as in the surface of the surface of the

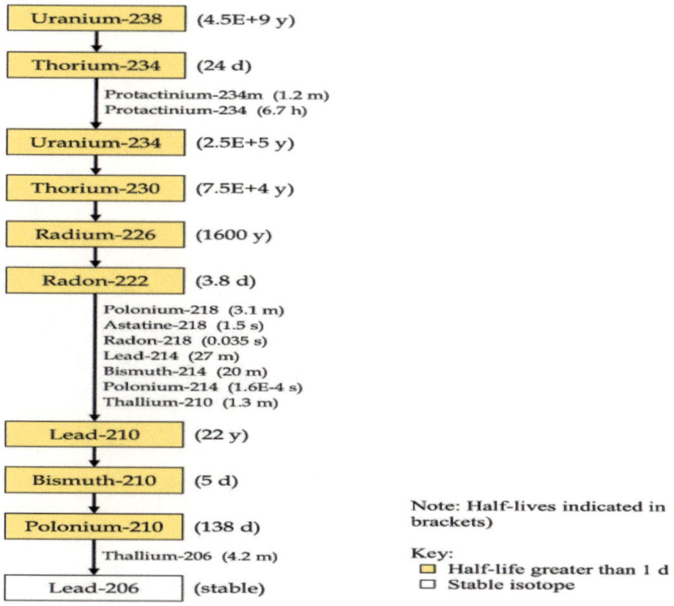

Figure 1.
The emission of radon from radium in underground rocks or soil to water [7].

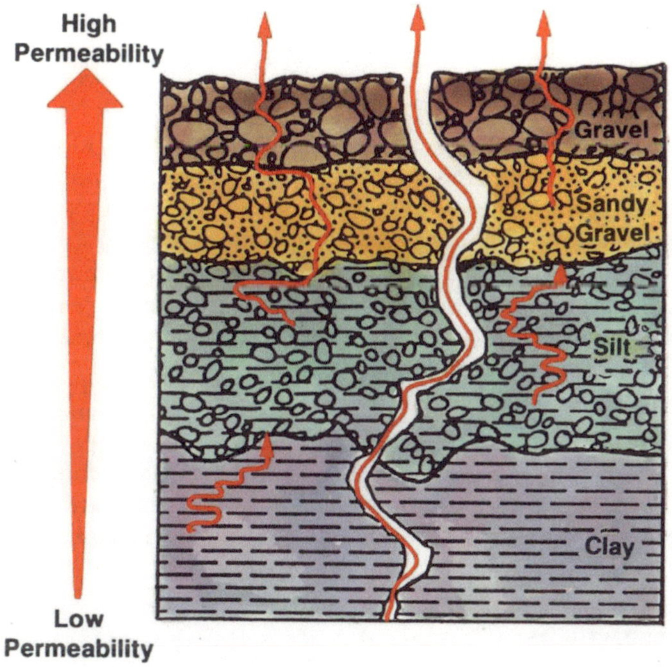

Figure 2.
Permeability of soil from underground to surface soil [9].

earth. Methods of obtaining water from underneath the earth in wells or boreholes disperse high levels of radon to the atmosphere [12]. **Figure 3** illustrates the dispersion of radon from rocks to water.

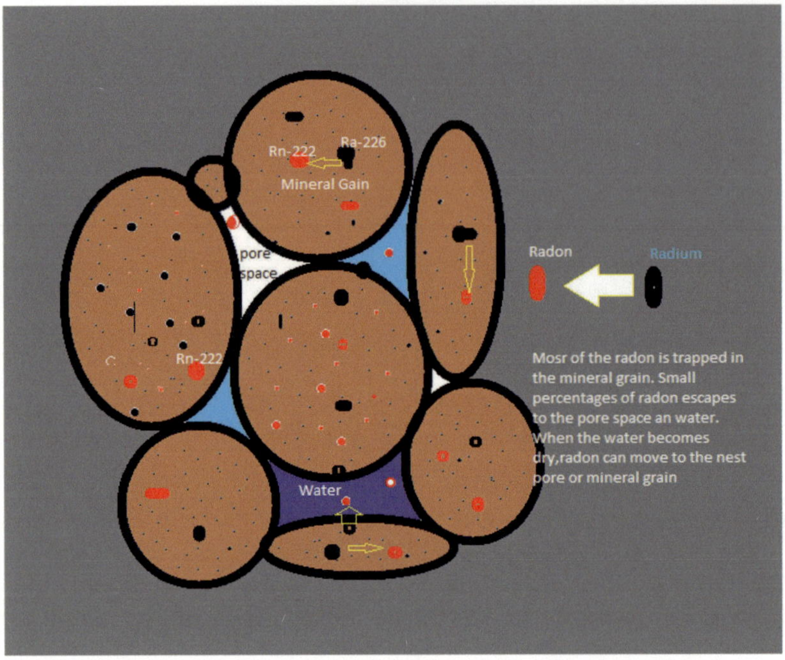

Figure 3.
Migration of radon in underground radium rocks to water [7].

1.3 Mitigation systems

Radon mitigation system is a practical approach to minimize high levels of radon emanation in identified areas through determining radon entry points and the nature of the foundation of buildings. In mines, to deal with radon entry points, radon sealants and bulkheads are used to block or control radon emanation from rocks, while in dusty atmospheres of the mine dust, controls such as air filters are of use [13] . In communities closer to radon-prone areas, natural ventilation system is important followed by radon suction system to drive radon away from the house or building.

1.4 Aim

This work presents an overview on radon emanation measurements and the possibility of applying mitigation system to reduce the public exposure to radon, which will in turn improve the public health and allow for better land use.

2. Radon measurement

Radon is measured by a wide range of techniques developed over years and improved in sensitivity, portability, and performance parameters [14]. The radon measurement technique mostly depends on the concentration of radon-222 and the characteristics used to classify them as follows:

- whether the technique measures radon-222 and radon progeny

- detection of radioactive decay mode—whether alpha, beta, or gamma
- time resolution

An alpha particle detected can be used to measure radon-222 decay products by using scintillation counters. Conversely, in the measurement of radon-222, gamma rays emitted from radon-222 daughters can be used to determine or measure radon-222, for example, bismuth-214 and lead-214 [15].

Time resolution sampling analysis and measurement are classified into three techniques, namely grab sampling, continuous technique, and integrating technique. Integrating technique utilizes passive detectors to measure the integrated radon-222 concentration annually or monthly mostly in buildings. Continuous technique provides the simultaneous act of sampling and determination of radon-222 in samples of soil or water in minutes, hours, and days with a device such as scintillation radon monitors and smart radon duo. Grab sampling technique includes the collection of samples of groundwater air samples, in a short time whereby devices such as RAD7 are used to measure radon-222 in samples collected. Awhida [16] states that radon measurement technique differs to the point that one technique cannot meet the requirement such as radon survey type, environmental parameter measurement, and cost of the apparatus.

2.1 Determination of radon-222 by its progeny

Baskaran [17] states that the decay of ^{222}Rn leads to the formation of numerous radon-222 progenies before ^{210}Pb:

$$^{222}\text{Rn (3.82)} \rightarrow {}^{218}\text{Po(3.10 min)} \rightarrow {}^{214}\text{Pb(26.8 min)} \rightarrow {}^{214}\text{Bi(19.7 min)} \rightarrow$$
$$^{214}\text{Po(10.2 min)} \rightarrow {}^{210}\text{Pb(22.3 yr)} \rightarrow {}^{210}\text{Bi(5.0d)} \rightarrow {}^{210}\text{Po (138.4d)}.$$

All the progenies of ^{222}Rn are metals, and the longest-lived progeny is ^{214}Pb. Polonium(Pb) and bismuth (Bi) are particle reactive and attach to atmospheric aerosol particles. ^{222}Rn concentration can be determined from the activity concentration of its progenies. Using Geiger Muller counter, ^{222}Rn activity concentration can be determined from the alpha particles of progenies such as ^{218}Po (6.002 MeV), ^{214}Po (7.687 MeV) or 214Bi (β_{max} = 3.272 MeV), and ^{214}Pb (β_{max} = 1.022 MeV) beta decay particles.

2.1.1 ^{222}Rn determination by solid-surface barrier detector

One of the ways of determining ^{222}Rn activity it is through alpha spectrometry by counting alpha emitting progenies of ^{222}Rn.

A known quantity of radon air is drawn by two filters in a cylindrical system, whereby one filter keeps the radon progeny and particular matter, the other filter receives radon. On the second filter, ingrowth of radon takes place, the second filter faces the surface barrier detector, and therefore, alpha spectra are produced.

In a simple method, positively charged progenies are collected onto a metal surface that is placed inside chamber's negative potential. For example, the positively charged 218Po particles gathered on a metal surface are counted by a surface barrier detector.

The advantage of using a surface-barrier detector is high resolution, low background, ability to distinguish the signals of ^{222}Rn progenies and ^{222}Rn due to the fact that they have varying alpha energies [17].

Type	Size	Accommodation capacity	Cost
Accumulation	Normally small in size (0.2–1.3	Accommodates limited number of radon detectors	Cost less to build
Flow-Through, Walk in chamber	Normally big in size	Accommodates a lot of radon detectors	Expensive to build

Table 1.
Radon chamber.

2.1.2 ^{222}Rn determination using beta counter

Assuming that there is secular equilibrium between ^{222}Rn and its progeny, the beta particles of ^{214}Pb and ^{124}Bi on a filter can be used for determining the radon concentration by measuring their beta activity. Beta counter such as Geiger Muller counter is used to count the radon progeny on filter paper. Also, plastic scintillators installed on photomultipliers tubes can be used to determine activity.

2.1.3 Direct progeny monitoring technique

One of the ways to measure a time integrated radon progeny for determination of radon is with direct radon progeny sensor, which is made up of solid-state nuclear track detector fitted with absorber of sufficient thickness. The direct radon progeny absorber comprises aluminized Mylar and cellulose nitrate appropriate thickness of 37 cm, whereby 7.67 MeV alpha particles emitted from ^{214}Po are detected [18].

2.1.4 Radon chamber

A radon chamber is a container that houses a radon atmosphere for calibration of radon detectors and conduction of studies in the field of radon metrology. It consists of tools for relative humidity, pressure, and temperature, which must be at a constant range for calibration of radon detectors. **Table 1** summarizes the type, size, and accommodation capacity of radon chambers; furthermore, the radon generating source must be traceable to an international or national standard.

3. Studies of indoor radon emanation levels in South Africa and possible mitigation system

Over the years, mining activities and mine dumps without proper regulation or disposal have led to the rise in background activity due to the presence of naturally occurring radioactive materials (NORM). Naturally occurring radioactive materials are in varying activities in soil, rocks, and underground water. In mining activities, when NORMs are processed, their activity becomes increased and is known as TENORM. NORMs contain radionuclides from the decay of uranium-238, thorium-232, and potassium-40, which result in the formation of radium and radon, posing a health hazard to exposed workers or nearby communities.

Many of the gold mines in South Africa process NORMs, and there is lot of mine waste, of which some is closer to communities in the Gauteng province [19]. There are abandoned mines of which some communities or instance Tudor-shaft houses build on top of mine tailings. An intervention of government and regulatory body in terms of regulating mines especially waste and dust control is necessary for the lessening of health threat by radioactive materials from contaminated sites. Maximizing distance between the mine and communities and restricting access to abandoned mine dumps or tailing dams help in preventing NORMs from reaching dwellings.

3.1 Radon emanation mitigation system

Radon mitigation system is a set of steps designed with a goal to achieve the reduction of radon in buildings in order to minimize the risk of lung cancer as seen in **Figure 4**. USEPA [2] recommends that radon levels of 148 Bq/m^3 (4 pc/l) or above in buildings must be reduced by radon mitigation system, also radon levels between 74 Bq/m^3 (2 pc/l) and 148 Bq/m^3 (4 pc/l) are to be considered for radon reduction, as no level of radon is safe.

For the application of radon mitigation systems on houses, how the house is constructed in terms of its foundation, as seen in **Figure 5**, determines the mitigation system. There are mainly three types of house foundation, namely:

- slab on grade (concrete poured at ground level),
- basement and
- crawlspace (a shallow unfinished space under the first floor)

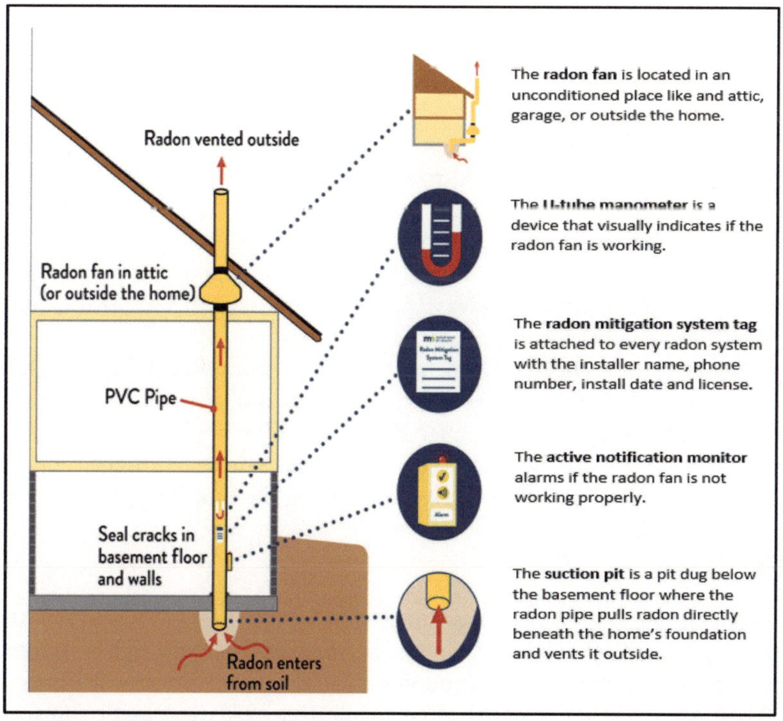

Figure 4.
Radon mitigation process [20].

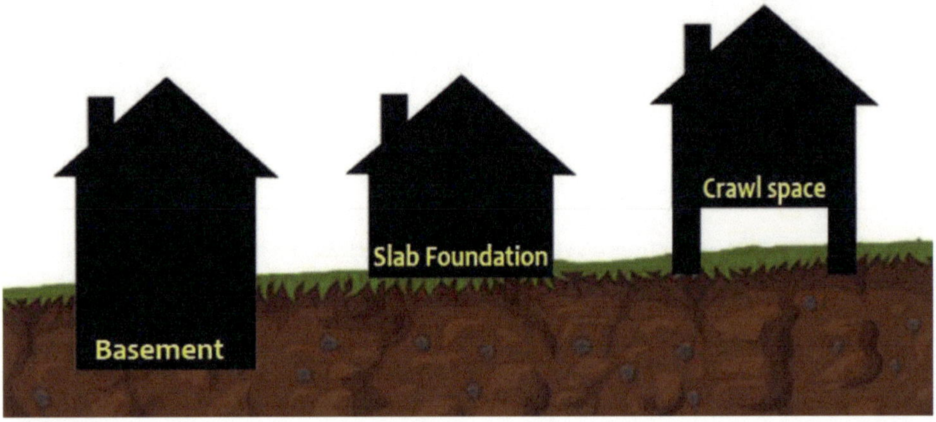

Figure 5.
Foundation types [21].

USEPA [2] recommends radon reduction methods that prevent radon from entering the house rather than a method that deals with radon when it has entered the house. A Radon Specialist runs diagnostic test to check possible pathways where radon emanates by shooting chemical smoke into cracks, holes, or drains and observing possible emission points. Then mitigation methods are followed.

3.1.1 Basement or a slab-on-grade foundation

For house foundations that are basement or a slab-on-grade foundation type, a radon mitigation system applicable is commonly one of four types of soil suction: active or passive sub-slab suction, drain-tile suction, block-wall suction, or sump-hole suction [2]. The sub-slab suction also known as sub-slab depressurization is a commonly used system whereby holes are drilled on the foundation of the house to beneath the foundation (to crushed rocks or soil), for insertion of suction pipes. Number of suction pipes installed depends on how radon air can be extracted from beneath the house concrete or slab with the use of the radon vent fan connected to the suction pipes to the air outside. Drain tiles or perforated pipes are commonly used in some homes for directing water away from the foundation. Therefore, suction on these pipes or tiles can effectively reduce radon levels.

Block wall suction is a radon mitigation system that is good for houses with basement where their walls are made of hollow blocks. The block wall suction system consists of a fan and ductwork, radon air is drawn from the hollow blocks of the basement and vented outside through exhaust fan. The block wall suction system is advantageous over other mitigation system as it prevents radon from reentering the building [22].

3.1.2 Crawlspace

Sub-membrane suction system is a mitigation system ideal for houses without basement. A high-density polyethylene plastic sheet is used as a radon reduction barrier by covering the exposed dirt or soil on the floor and also the walls as seen in **Figure 6**. Then, suction pipes are installed through the plastic sheet to depressurize the soil and draw the radon gas outside the house through installed radon fan and suction pipe [23].

Figure 6.
Example of sub-membrane suction system [23].

Other types of radon mitigation systems applicable to any house foundations include sealing, natural ventilation, house or room pressurization [2].

The sealing of cracks and openings method to is the primary part of most radon mitigation systems. Identification of areas where there is radon entry is mostly not easily detectable as cracks happen overtime. Sealing of cracks or voids minimizes the flow of radon into the house.

House or room pressurization system keeps radon air trapped in the basement by the blowing of the air into the basement. The limitations of the technique affecting its effectiveness are the house construction, appliances in the house or house occupants' lifestyle.

Natural ventilation happens in all houses. Ventilation for radon reduction can be improved by opening doors, windows, and vents on lower floors. When windows and doors are closed, radon concentration returns to its previous value in about 12 hours [2].

3.2 High indoor radon levels in South Africa and applicable radon mitigation system

3.2.1 Gauteng Province

In South Africa, there are studies of indoor radon measurements, some of which indicate levels of radon higher than 148 Bq/m^3 recommended by EPA [24]. In a study done by Radebe [25] for the design of radon chamber, alphaguards that were calibrated measured radon levels above 1000 Bq/m^3 from Tudor shaft soil samples.

Tudor shaft, an informal settlement in Krugersdorp, is known to be affected by gold mine shaft, and tailing dam is one of the areas of note that radon mitigation methods must be implemented. The inhabitants of Tudor shaft built their shacks on top of mine dump with radioactive uranium soil as seen in **Figure 7**. Vegetables are

Figure 7.
(a) Example of shack interior for Tudor shaft. (b) Children playing on soil that has potentially higher levels of radiation. (c) Shacks built on soil that potentially have higher levels of radiation [26, 27].

grown on top of soil potentially carry high levels of radioactivity and environment due to polluted air from mine tailings. The community is at risk of radioactive health risk arising from mine dumps. Typical entry points in shacks in Tudor shaft comprise cracks on the floor or ground, holes on mats used to cover the soil, and gaps found on the edges of the interior of the shack between the mat and the shack or poles. It is also noted that when doors and windows are closed, radon accumulates and returns to an average value, while during the day, it reduces indoors.

A radon mitigation system applicable to Tudor-shaft inhabitants for shacks can be a passive sub-slab depressurization system. A passive sub-slab system relies on stack effect, which is a term that defines radon reduction by the reliance on air pressure differentials to extract radon from underneath the foundation to the outdoor air via vent pipes. Mats or floor covers must be checked to any opening and removed for the installation of a high-density polyethylene plastic to retain the radon gas. Then the radon vent pipe can be installed for moving the radon retained by the plastic to the outdoor air as seen in **Figure 8**. Furthermore, when a floor cover or mat is placed on top of the high-density polyethylene plastic, openings or voids must be sealed with caulk or epoxy sealant. Passive sub-slab system together with sealing and natural ventilation application can reduce the radon in informal settlement such as Tudor shaft especially for shacks without electricity means. The government has done a good job in relocating some families affected by mine dumps.

A best solution would be to relocate the community to a better place. A temporary solution would be to apply radon mitigation methods to minimize the risk of lung cancer, also community awareness about reducing emanation of radon via natural ventilation and sealing of openings or cracks is important.

DOI: http://dx.doi.org/10.5772/intechopen.109065

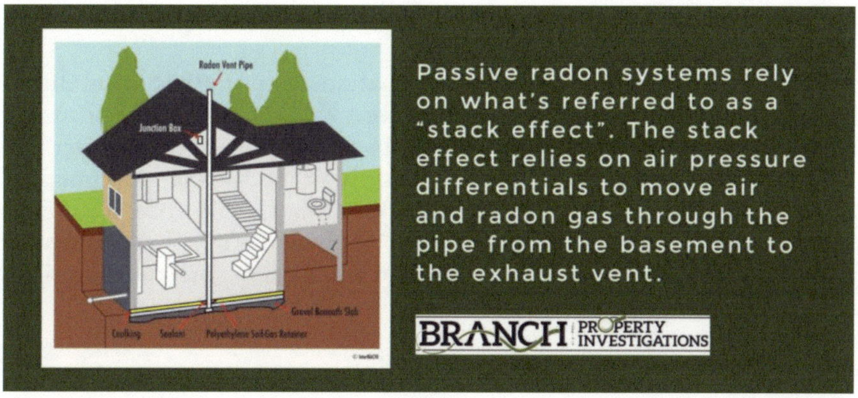

Figure 8.
Passive sub-slab system [28].

3.2.2 Western cape

Houses in Paarl, Western Cape province, with a type of crawlspace foundation used for storage by occupants recorded radon levels off up to 800 Bq/m^3 as seen in **Figure 9** [30]. Also houses with wooden floor recorded higher radon levels than concrete floor. At the foothill off Paarl Mountain, higher levels of radon levels were found and some of the houses are nearby the foothill. An active sub-slab depressurization system and sub-membrane suction system for crawl space can be applied for houses at Paarl.

Furthermore, methods of foundation crack repair and staples are essential for the reduction of radon in houses, although mitigation systems are the best. Foundation crack repair inhibits the entrance of radon indoors. By utilization of products such as concrete staples or epoxy, which works like a glue, and holes or cracks that are in the concrete or wooden foundation are repaired and reinforced [31], which thus stops the primary entrance of radon in houses.

The houses with walls having stability problems tend to lean or bow; therefore, carbon staples can be used to repair and reinforce the cracked or bowed wall to achieve radon emanation reduction. Epoxy, which comprises epoxy resin and hardener, is essential for closing the cracks in foundation where there is radon entry [31] .

Figure 9.
Radon in Paarl crawlspace type of a foundation [29].

4. Conclusion and recommendation

There are areas of high levels of radon emanation in South Africa. and there are mitigation techniques to be applied. Therefore, there is a need of trained radon mitigation specialist to precisely reduce radon in radon-prone areas to lower levels. This will prevent rates of lung cancer from going high and thus support good public health. In addition, radon calibration facilities play a major role in determining accurate measurement of radon by radon detectors. Implementation of radon awareness campaigns that cover topics of radon mitigation is necessary with collaboration with the national regulatory body to affected communities.

Author details

Moses Radebe[1,2]* and Manny Mathuthu[1,2]

1 Center for Applied Radiation Science and Technology (CARST), North-West University (Mafikeng), Mmabatho, South Africa

2 National Research Foundation, South Africa

*Address all correspondence to: mozsmash@gmail.com

References

[1] Lubin JH, Boice JD, Edling C, Hornung RW, Howe GR, Kunz E, et al. Lung cancer in radon-exposed miners and estimation of risk from indoor exposure. Journal of the National Cancer Institute. 1995;**87**:817-827

[2] USEPA. Consumer's Guide to Radon Reduction. Washington DC: EPA Publication; Mar 2013[cited 2022 05 August 2022]; EPA 402/K-10/005:[20]. Available from: http://www.epa.gov/radon/radon/consumers-guide-radon-reduction-how-fix-your-home

[3] Fuente M et al. Performance of radon monitors in a purpose-built radon chamber. Journal of Radiological Protection. 2018;**38**(3):1111

[4] Speeman W, Pule J. South African Perspective for Radon in Dwellings and the Anticipated Regulatory Control Measures. in NNR Regulatory Information Conference. Pretoria, South Africa: National Nuclear Regulator; 2016

[5] Centomani GV, Di Carlo C, Lepore L, Remetti R. Design and commissioning of an innovative radon chamber with a single 226Ra source and continuous variation and control of concentration vs. time. Nuclear Instruments and Methods in Physics Research Section A: Accelerators, Spectrometers, Detectors and Associated Equipment. 2019;**940**:109-115

[6] Puukila S, Haigh P, Johnston A, Boreham DR, Hooker AM, Dixon D-L. A radon chamber specifically designed for environmentally relevant exposures of small animals. Journal of Environmental Radioactivity. 2020;**220-221**:106295

[7] Hararah SMAI. Investigation of Radon Pollution in Groundwater in the Southern Part of Gaza Strip - Palestine. Gaza: Master of Science in Water Resources Management, The Islamic University; 2007

[8] Hosoda M et al. Effect of soil moisture content on radon and thoron exhalation. Journal of Nuclear Science and Technology. 2007;**44**:664-672

[9] Gundersen LC et al. Geology of radon in the United States. Geological Social American Special Paper. 1992;**271**:1-16

[10] Smith JE, Cothern CR. Environmental Radon/Edited by C. Richard Cothern and James E. Smith, Jr. New York: Plenum Press; 1987

[11] Sokolov P, Sentenberg S. Heat Treatment Influence Evaluation on Radon-222 Emanation Factor in Building Material. International Research Journal. 2018;**6**:48-72

[12] Sukanya S, Noble J, Joseph S. Application of radon (222Rn) as an environmental tracer in hydrogeological and geological investigations: An overview. Chemosphere. 2022;**303**:135141

[13] Tukkaraja P, Bhargava R, Sridharan SJ. Radon in underground mines. In: Hammond A, B. D. B, Ashwath N, editors. Mining Technology. London, UK: IntechOpen; 2021

[14] Nagda NL. Radon: Prevalence, measurements, health risks and control. In: Nagda NL, editors. ASTM Manual Series: MNL 15. Philadelphia: American Society for Testing and Materials; 1994

[15] Elzain AEA. Radon monitoring in the environment. In: Adrovic F, editor. Radon. London, UK: IntechOpen; 2017

[16] Awhida A et al. Novel method of measurement of radon exhalation

from building materials. Journal of Environmental Radioactivity. 2016;**164**

[17] Baskaran M. Radon: A Tracer for Geological, Geophysical and Geochemical Studies. Springer Geochemistry, ed. Springer. Switzerland: Springer International Publishing; 2016

[18] Singh P, Singh P, Singh S, Sahoo BK, Sapra BK, Bajwa BS. A study of indoor radon, thoron and their progeny measurement in Tosham region Haryana, India. Journal of Radiation Research and Applied Sciences. 2015;**8**:226-233

[19] Kamunda C, Mathuthu M, Madhuku M. Health risk assessment of heavy metals in soils from Witwatersrand Gold Mining Basin, South Africa. International Journal of Environmental Research and Public Health. 2016;**13**(7):663

[20] Health, Minnesota Department. Radon Mitigation System. Minnesota. 2022 [Accessed 04 May 2022]; Available from: https://www.health.state.mn.us/communities/environment/air/radon/mitigationsystem.html

[21] Byhyu. Slab, Crawlspace or Basement? Do You Know What Type of Foundation to Choose—BYHYU 069 [Online]. 2017. Available from: https://www.byhyu.com/home--podcast/slab-crawlspace-or-basement-do-you-know-what-type-of-foundation-to-choose-byhyu-069 [Accessed 2022]

[22] Khan SM, Gomes J, Krewski DR. Radon interventions around the globe: A systematic review. Heliyon. 2019;**5**(5):e01737

[23] Defense R. Sub Membrane Ventilation [Online]. 2022. Available from: https://radondefenseva.com/radon-mitigation/sub-membrane-ventilation/ [Accessed 2022]

[24] Maheso AM. Radon Levels in South African Homes-Design Elements for a National Survey and Initial Results from Directed Sampling. Stellenbosch: Stellenbosch University; 2021

[25] Radebe M. The design of a radon chamber for the calibration of radon monitors at the Centre for Applied Radiation Science and Technology, Mafikeng, South Africa, in Centre of Applied Radiation Science and Technology. North-West University. 2019

[26] Environment. Revisiting the West Rand Goldfields in 2016 [Online]. 2016. Available from: https://www.environment.co.za/mining/wake-south-africas-abandoned-gold-mines.html

[27] Sowetan. Shack Dwellers Exposed to High-Risk Radiation [Online]. 2012. Available from: https://www.sowetanlive.co.za/news/2012-03-30-shack-dwellers-exposed-to-high-risk-radiation/ [Accessed June 20, 2022]

[28] Investigations BP. What is a Passive Radon Mitigation System? [Online]. 2020. Available from: https://branchinvestigations.com/what-is-a-passive-radon-mitigation-system/ [Accessed June 19, 2022]

[29] Lindsay R. Radon and other applied nuclear work at UWC. October 2016 [Accessed 05 August 2022]; Presentation of Radon Studies. Available from: http://www.nnr.co.za/wp-content/uploads/2016/10/2.-Pres-Radon-work-at-UWC.pdf

[30] Lindsay R, Newman R, Speelman W. A study of airborne radon levels in Paarl houses (South Africa) and associated source terms, using electret ion chambers and gamma-ray spectrometry. Applied Radiation and Isotopes. 2008;**66**:1611-1614

[31] Mitigation NR. Foundation Crack Repair and Staples [Online]. 2022. Available from: https://nordicradonmitigation.com/crack-repair-and-staples/

Section 2

Utilization and Resource Procurement

Chapter 8

Evaluation of Rare Earth Element Mine Sites for Environmental Impact

Michael T. Aide

Abstract

Rare earth elements are critical elements in the modern economy. Mining of rare earth elements has significantly intensified in the last several decades and studies of the environmental impact are in their infancy. In trace amounts, rare earth elements may support plant growth and development. At greater concentrations, rare earth elements are increasingly recognized as having a degree of mammalian toxicity; however, the mammalian toxicity potential may not be as acute as that for some heavy metals. The toxicity of rare earth elements requires detailed research to showcase toxicity thresholds for a wide range of ecosystem health. This study reveals case studies demonstrating that investigators rely on pollution indices, which do indicate that mining and ore processing possess environmental challenges. Further research has been identified to evaluate pollution indices for rare earth elements, especially concentrating on their biological availability.

Keywords: rare earth elements, pollution indices, mine contamination, human health impacts, environmental impact

1. Introduction

The rare earth elements (REE) are the 14 elements comprising the lanthanide series: cerium (Ce), praseodymium (Pr), neodymium (Nd), promethium (Pm), samarium (Sm), europium (Eu), gadolinium (Gd), terbium (Tb), dysprosium (Dy), holmium (Ho), erbium (Er), thulium (Tm), ytterbium (Yb), and lutetium (Lu) [1]. Lanthanum (La) is associated with rare earth elements because of its periodic table position. Frequently, scandium (Sc) and yttrium (Y) are grouped with the lanthanides given their similar chemical reactivity. The lanthanide series are elements characterized as having one or more electrons in the 4f electronic orbitals for their ground state configuration [1, 2]. Promethium is absent in the environment because promethium-145 (^{145}Pm) decays *via* electron capture into neodymium-145 (^{145}Nd), with a half-life of 17.7 years.

The typically trivalent rare earth elements have considerable ionic bonding character [1]. Cerium may have a valency of +3 or + 4, whereas europium may have a valency of +2 or + 3 [1]. The rare earth elements show a decrease in their ionic radii on progression from La to Lu, the so-called "lanthanide contraction". The "lanthanide contraction" supports greater chemical affinity for hydrolysis and chelate/complex stability with an increase in atomic number [1]. The LREE are the light rare earth elements, comprising of the elements from La to Eu, and the HREE are the heavy rare earth elements, comprising of the elements from Gd to Lu.

Rare earth element concentrations in soils, sediments, and other earth materials are dependent on their mineral assemblages, with rare earth element concentrations typically ranging from 0.1 to 100 mg kg^{-1}. Scandium concentrations across the earth's crust are in the range from 16 to 30 mg Sc kg^{-1}, with mafic and argillaceous materials showing greater scandium concentrations. Typical soil scandium concentrations are from 0.8 to 28 mg Sc kg^{-1}. Yttrium concentrations average from 20 to 30 mg Y kg^{-1} across crustal materials, where yttrium concentrations in soil range from 7 to 200 mg Y kg^{-1} [3]. The Oddon-Harkins rule states that an element with an even atomic number has a greater concentration than the next element in the periodic table. The rare earth elements typically follow the Oddon-Harkin rule. The Post-Archean Australian Average Shale (PAAS), North American Shale Composite (NASC), and Upper Continental Crust (UCC) reflect the Oddon-Harkin rule (**Table 1**).

Kabata-Pendias [3] has compiled many studies observing REE abundances in mafic and felsic igneous rocks, sedimentary rocks (argillaceous, sandstones, and calcareous), and soils. Kabata-Pendias also compiled reference data on terrestrial plant species. Commonly occurring REE-bearing minerals include: (i) fluorite (Ce), (ii) allanite (Ce), (iii) sphene (REE), (iv) zircon (HREE), (v) apatite (LREE), (vi) monazite (Ce and La), (vii) xenotime (REE), (viii) rhabdophane (Ce and REE), and

Element	PAAS	NASC	K-P	UCC
		mg/kg		
La	38.2	32	26.1	3.1
Ce	79.6	73	48.7	63
Pr	8.83	7.9	7.6	7.1
Nd	33.9	33	19.5	27
Sm	5.55	5.7	4.8	4.7
Eu	1.08	1.24	1.2	1.0
Gd	4.66	5.2	6.0	4.0
Tb	0.774	0.85	0.7	0.7
Dy	4.68	5.8	3.7	3.9
Ho	0.991	1.04	1.1	0.83
Er	2.85	3.4	1.6	2.3
Tm	0.405	0.5	0.5	0.3
Yb	2.82	3.1	2.1	2.0
Lu	0.433	0.48	0.3	.31

PAAS is Post-Archean Australian Average Shale, NASC is North American Shale Composite, and UCC is the upper continental crust. PAAS and NASC reported in McLennan [4, 5], K-P reported in Kabata-Pendias [3], and UCC reported in Rudnick and Gao [6].

Table 1.
REE abundances for various parent materials.

(ix) bastnaesite (REE) [4]. Mineral and soil assemblies typically show greater LREE concentrations than HREE; however, the LREE/HREE ratio may vary with zircon abundances. In a review, Van Gosen et al. [7] noted that many of the significant rare earth element deposits occur in carbonatites (carbonate igneous rocks). Peralkaline igneous systems, magmatic-magnetite–hematite bodies, and mafic gneiss-bearing xenotime-monazite deposits are also important sources for rare earth element extraction.

Van Gosen et al. [7], U.S. geological survey [8], Ramos et al. [9], Kim and Jariwala [10], and Van Veen and Melton [11] review the rare earth element available supply and known resources, concentrating on nations having substantial reserves: Australia, China, India, Malaysia, Russia, and the United States. China and Brazil have significant rare earth element ore extractions.

2. Rare earth elements and the modern economy

Rare earth elements are critical to scientific and industrial advancement, ranging from energy to information technologies. Some rare earth elements have unique applications, such as gadolinium as contrast agent essential for magnetic resonance imaging (**Table 2**). Other important rare earth element uses include (i) permanent magnets (neodymium-iron-boron), (ii) cell phones, (iii) electric motors for electric cars, (iv) steel making, and (v) phosphors (yttrium, cerium, lanthanum, europium, and terbium) Van Gosen et al. [7]. Additional applications may utilize an array of rare earth elements, such as yttrium, europium, dysprosium, and holmium, in the manufacture of lasers (**Table 2**) [7, 10].

3. Rare earth elements and their influence on living organisms, including human health

Kabata-Pendias [3] documented rare earth element influences on plant physiology, noting several studies, reporting that the REE stimulates seed germination, root growth, nutrient uptake, biological nitrogen fixation, chlorophyll synthesis, and photosynthesis. Kabata-Pendias was careful to note that further research is required for conformation and whether the rare earth elements are considered plant essential elements. The influence of rare earth elements, because of their production, processing, and usage on mammalian health, including human health, has not been widely investigated [12]. Of the elements comprising the lanthanide series and their influence on human health, the elements Ce, La, Gd, and Nd have received the most scrutiny [13]. Compounding rare earth element research is that many chemicals involved in rare earth element mining, recovery, primary and secondary processing, and recycling are involved; thus, it is difficult to isolate the influence of rare earth elements on human health.

Industry and occupational health research are limited because ore extraction and refining are localized in only a few nations [12, 13]. Bioaccumulation of rare earth elements appears to be restricted near mine and ore processing sites. There is growing evidence of the adverse effect of gadolinium (Gd) on skin conditions and nephrogenic systemic fibrosis because of Gd's use as an element in contrast agents used in magnetic resonance imaging [13]. Short-term exposure animal studies suggest rare earth element toxicity involves the liver, lungs, blood, and nervous systems. The global use of

Element	Usage
Scandium	Aluminum alloys, aerospace components
Yttrium	Lasers, computer displays, microwave filters
Lanthanum	Oil refining, hybrid-car batteries
Cerium	Oil refining, catalytic converters, and lens production
Praseodymium	Aircraft engines, carbon arc lights
Neodymium	Computer hard drives, cell phones, and high-power magnets
Promethium	Portable X-ray machines, nuclear batteries
Samarium	High power magnets
Europium	Lasers, computer displays, and optical electronics
Gadolinium	Magnetic resonance imaging contrast agents
Terbium	Solid-state electronics, sonar systems
Dysprosium	Lasers, high-power magnets, and nuclear reactor control rods
Holmium	Lasers, high-power magnets
Erbium	Fiber optics, nuclear reactor control rods
Thulium	X-ray machines, superconductors
Ytterbium	Portable X-ray machines, lasers
Lutetium	LED lightbulbs

Source: Van Gosen et al. [7], Kim and Jariwala [10].

Table 2.
Rare earth element usage in modern industries and technologies.

cerium oxide nanoparticles as a catalytic additive in diesel fuel may be an air and soil pollutant. Conversely, rare earth elements may confer beneficial antioxidant activity. In an extensive review, Rim [12] noted that the rare earth elements high redox potential supported oxidative stress, which may enhance diabetes, atherosclerosis, inflammatory conditions, high blood pressure, neurodegenerative diseases, and cancer. Rim [12] further noted that selective chemicals used for rare earth element ore extraction, processing, and manufacturing may contribute to the environmental impact. The combined toxicities of the rare earth elements may influence soil pH and influence the human toxic response.

Many soil studies have been conducted to determine if soil rare earth element concentrations may have sufficient soil variability to influence environmental responses. In Brazil, Landim et al. [14] established soil rare earth element quality references and assessed their spatial distributions. The mean background concentrations in soils followed the abundance of the earth's upper crust: Ce > La > Nd > Pr > Sm > Dy > Gd > Er > Yb > Eu > Tb > Lu. In the Piaui state in Brazil, the ∑REEs across the mesoregions were (i) southeast (263 mg kg^{-1}), (ii) north and central-north (90 mg kg^{-1}), and (iii) southwest (40 mg kg^{-1}) [14].

The effects of rare earth elements on aquatic biota are largely unknown [15, 16]. Gonzalez et al. [15] researched the sensitivity of aquatic organisms to REE's. Lanthanide series toxicity increased with atomic number for *Aliivibrio fischeri* and *Pseudokirchneriella subcapitata*, whereas the lanthanide series sensitivity of *Daphnia magna*, *Heterocypris incongruens*, *Brachionus calyciflorus,* and *Hydra attenuata* were effectively equal across atomic number [15]. Comparison of environmental risk indexes indicated that lanthanide concentrations immediately downstream of wastewater treatment plants are most problematic. Malhotra et al. [16] reviewed previous studies to understand the bioavailability, accumulation, interaction, and toxicity criteria of rare earth elements with aquatic organisms. Previous studies demonstrated that rare earth elements frequently provide a stimulatory influence at low dosage, whereas adverse influences are observed at higher dosages.

In Canada, the exposure-response relationships of three native plant species (*Asclepias syriaca*, *Desmodium canadense*, and *Panicum virgatum*) and two commonly used crop species (*Raphanus sativus*, and *Solanum lycopersicum* L.) to lanthanum, yttrium, and cerium were evaluated [17]. Germination was influenced by Ce at lower pH values. All species showed growth inhibition with Ce addition. Growth inhibition was evident for *A. syriaca* and *D. canadense* with rare earth element addition [17].

Li, et al. [18] assessed the toxicity of lanthanum, after 3 to 4 weeks of exposure, to five representative soil invertebrates. Toxicity was related to (i) total lanthanum, (ii) 0.01 M $CaCl_2$-extractable lanthanum, and (iii) porewater lanthanum concentrations. Reduced growth of Isopod (*Porcellio scaber*) showed the most sensitive response. Reproduction issues involving earthworms (*Eisenia andrei*), enchytraeids (*Enchytraeus crypticus*), springtails (*Folsomia candida*), and oribatid mites (*Oppia nitens*) were observed at various elevated soil La concentrations. This study suggested that La may affect soil ecosystems at La concentrations slightly above natural background levels (6.6–50 mg La kg^{-1}).

In China, Zhou et al. [19] performed experiments involving dry grass landfilling, chicken manure broadcasting, and plant cultivation to reclaim a rare earth element mine. After 2 years of restoration, soil organic matter, available potassium, available phosphorus, and acid phosphatase activity were improved. Soil physical properties (bulk density, water holding capacity, pH, and electrical conductivity), nutrient availabilities, and enzyme activities after 5 years were either similar or less impacted than soil not impacted by rare earth mining activities.

4. Commonly used pollution indices

Pollution indices are calculated estimates of the degree of soil and sediment contamination, usually associated with heavy metals. Pollution indices are also employed to (i) assess soil quality and health, (ii) predict ecosystem sustainability, and (iii) discriminate between natural processes and anthropogenic processes to explain heavy metal distributions in soil profiles [20, 21]. Many of the pollution indices rely upon a proper selection of a geochemical background (GeoBase) [20, 21]. The geochemical background, if properly selected, will permit an estimation of the intensity of the heavy metal or rare earth element pollution.

The geochemical background (GeoBase) for an element estimates the natural variations in concentration in the surficial environment [22] or is a "measure that is used to differentiate between the concentration of the natural compound and the concentrations with an anthropogenic influence in a given environmental sample" [23]. Kowalska et al. [21] assessed 18 pollution indices and estimated their strength and weaknesses. Diwa [24], Elvira [25], Lawrence et al. [26], Barbieri et al. [27], Ghrefat et al. [28], and Gargouri et al. [29] provided additional information on interpreting the environmental impact of pollution indices values. In the Czech Republic, Weissmannová et al. [30] chronicled the potential ecological risk and human health risk assessment in soils influenced by coal mining and metal processing.

Indices	Description
Geoaccumulation index (Igeo)	Igeo = $\log_2$[HMconc/(1.5 x GeoBase)], where HMconc is the concentration and GeoBase is the geochemical reference concentration.
Single pollution index (PI)	PI = HMconc/GeoBase
Enrichment factor (EF)	EF = [HMconc/LV]/GeoBase/LV], where LV (low variability reference) is the element concentration considered as not supplied or depleted.
Contamination factor (CF)	CF = sample mean concentration relative to preindustrial concentration.
Biogeochemical index (BGCI)	BGCI = soil metal concentration of O horizon relative to A horizon.
Sum of contamination (PIsum)	PIsum = $\sum$ PI, where each PI is PI = HMconc/GeoBase.
Nemerow pollution index (PInem)	PInem = $\{[(1/n)\sum PI)^2 + PImax^2]/n\}^{0.5}$, where n number of metals sampled and PImax is the maximum PI value.
Pollution load index (PLI)	PLI = $\{\prod PI\}^{1/n}$ (the harmonic mean of the PI's
Average single pollution index	PIaverage = $(1/n) \sum$PI.
The vector modulus of pollution	PIvector = $\{(1/n)\sum PI^2\}^{0.5}$.
Multi-element contamination	MEC = $\{\sum$(HMconc$\}$/n.
Degree of contamination	Cdeg = $\sum$CF (multiple elements)
Potential ecological risk factor	Eir = Ti x HMconc/GeoBase, where Ti is a factor of a particular heavy metal's toxic reaction.
Hazard quotient (HQ)	HQ = element concentration/element concentration where no environmental effect was observed.

Table 3.
Frequently employed pollution indices.

Frequently, employed pollution indices are briefly described in **Table 3** and more fully described below.

The geoaccumulation index (Igeo) is determined from the elemental concentration and the GeoBase. The GeoBase is the concentration value of the element selected from a geochemical reference background, and it is critical to select an appropriate geochemical background, with various research investigations using PAAS, NASC, UCC, or local baselines perceived as preindustrial or non-impacted. The Igeo value and the corresponding pollution level is (i) less than or equal to 0 is not impacted, (ii) 0 to 1 is at most moderately impacted, (iii) 1 to 2 is moderately impacted, (iv) 2 to 3 is moderate to highly impacted, (v) 3 to 4 is highly impacted, (vi) 4 to 5 is high to very highly impacted, and (vii) 6 is very highly impacted. The single pollution index evaluates the degree of heavy metal or rare earth element accumulation in soil or sediment relative to a reference GeoBase. The single pollution index estimates the total amount of an element's accumulation and does not indicate the bioavailability of the heavy metals. The single pollution index (PI) allows the comparison of sites or different soils over time. The enrichment factor (EF) estimates the heavy metal's anthropogenic impact and is determined as EF = [HMconc/LV] of sample/GeoBase/LV] of background, where LV (low variability reference) is the selected element concentration considered as not supplied or depleted. Typically, Fe, Al, Ca, Ti, Sc, and Mn have been used as the reference element. If the EF value ranges from 0.5 to 1.5, the likelihood of anthropogenic activity is low. The selection of the GeoBase that reduces the metal variability is critical for assessment. If the EF value and the enrichment level are: (i) less than 1, there is no likelihood of element enrichment (impact), (ii) 1 to 3 is minor enrichment, (iii) 3 to 5 is moderate enrichment, (iv) 5 to 10 is moderately severe enrichment, (v) 10 to 25 is severe enrichment, (vi) 25 to 50 is very severe enrichment, and (vii) more than 50 is extremely severe enrichment.

The contamination factor (CF) estimates the preindustrial increase of heavy metals or rare earth elements in soil and sediment. CF is estimated as the mean of more than five samples for a particular heavy metal or rare earth element relative to the heavy metal concentration from preindustrial samples. The selection and evaluation of the quality of the preindustrial reference samples may be difficult. Appropriate ratings are: (i) less than 1 is low contamination (impact), (ii) 1 to 3 is moderate contamination, (iii) 3 to 6 is considerable contamination, and (iv) greater than 6 is considered high contamination [25]. The biogeochemical index (BGCI) is estimated as a ratio of the heavy metal or rare earth element concentration in the soil's O horizon to that element's concentration in the soil's A horizon. If the BGCI has values greater than 1, then there exists an increased heavy metal or rare earth element adsorption in the O horizon. The BGCI is well suited for forest soils; however, the biogeochemical index lacks consideration of the heavy metal biological availability.

The sum of contamination (PIsum) involves a suite of heavy metals or rare earth elements and is simply the sum of the individual PI values for the sampled heavy metals. The calculation of the PIsum must contain all of the relevant heavy metals. The selection of the geochemical database must be appropriate to the site assessment, typically employing local and preindustrial sampling. Similar to the BGCI, the Pisum does not consider heavy metal biological availability. The Nemerow pollution index (PInem) is defined as $\text{PInem} = \{[(1/n)\sum \text{PI})^2 + \text{PImax}^2]/n\}^{0.5}$, where n is the number of heavy metals sampled, and PImax is the maximum PI value for all of the heavy metals. The usage of the appropriate geochemical database, baseline values, or threshold levels must be ascertained. The Nemerow pollution index directly reflects the soil or sediment environmental pollution and highlights the heavy metal having the greatest environmental presence or intensity of pollution.

The pollution load index (PLI) is calculated as $\{\prod \text{PI}\}^{1/n}$ and is simply the harmonic mean of the PI's for the analyzed heavy metals or rare earth elements. The pollution load index does not consider heavy metal biological availability and is based on the reliability of the PI values. The average single pollution index (PIaverage) is estimated as the average of the individual PI values. The average single pollution index does not consider heavy metal or rare earth element biological availability and is based on the reliability of the PI values. Appropriate ratings are: (i) less than 1.5 is very low contamination, (ii) 1.5 to 2 is low contamination, (iii) 2 to 4 is moderate contamination, (iv) 4 to 8 is high contamination, (v) 8 to 16 is considered very high contamination, and (vi) 16 or greater is considered extreme contamination [25]. The vector modulus of pollution index (PIvector) is estimated $\{(1/n)\sum \text{PI}^2\}^{0.5}$. The vector modulus of pollution index does not consider heavy metal or rare earth element biological availability and is based on the reliability of the PI values.

The multi-element contamination index (MEC) is estimated as $\text{MEC} = \{\sum(\text{HMconc of element i/tolerable level for element i})\}/n$. The multi-element contamination index does not require an assessment of the variation in natural processes. The degree of contamination (Cdeg) is estimated as $\text{Cdeg} = \sum \text{CF}$. The degree of contamination does not require an assessment of the variation in natural processes.

The potential ecological risk factor (Eir) is estimated as Eir from an estimate of the element's toxicity (Ti factor) and associated PI value, where the Ti has values of 1 for zinc (Zn), 2 for chromium (Cr), 5 for Nickel (Ni), Copper (Cu) and lead (Pb), 10 for arsenic (As), and 30 for cadmium (Cd). Ti values for the rare earth elements are not yet determined. The potential ecological hazard index (RI) is estimated as $\text{IR} = \sum \text{Eir}$.

If the Eir risk values are: (i) less than 40, which implies low ecological risk, (ii) 40 to 80 implies moderate ecological risk, (iii) 80 to 160 implies appreciable ecological risk, (iv) 160 to 320 implies high ecological risk, and (v) more than 320 implies series ecological risk. If the IR risk values are: (i) less than 150, which implies low ecological risk, (ii) 150 to 300 implies moderate ecological risk, (iii) 300 to 600 implies high ecological risk, and (iv) more than 600 implies series ecological risk [24]. The hazard quotient (HQ) of a rare earth element is estimated as the rare earth element concentration relative to the rare earth element concentration, where no environmental effect was observed.

5. Case studies of rare earth elements environmental impact because of mining, processing, and utilization

Rare earth element mines differ from other mines, that is, rare earth element ores are not highly concentrated, and the quantity of tailings is frequently substantial compared to the quantity of rare earth elements recovered. Krasavtseva et al. [31] investigated heavy metals and rare earth elements at mine sites in the Kola Subarctic (Russia). Noting that finely dispersed materials from mine tailings exhibited from 1.5 to 3 times the heavy metal and rare earth element concentrations, they calculated pollution assessments using the (i) geoaccumulation index, (ii) enrichment factor, (iii) potential ecological risk factor, and the (iv) potential environmental hazard index. Increased mobilized heavy metal and rare earth elements were observed for mine tailing leachates. The heavy metal and rare earth element leachate concentrations were increased with (i) reduced pH levels, (ii) elevated organic carbon levels, and (iii) increased temperatures.

In China, Bai et al. [32] investigated six rare earth element mines for their environmental impact and conducted a cross-sectional comparison. Accurate resource and environmental carrying capacity (RECC) assessments are critical for ensuring that rare earth element exploration and extraction activities are appropriate and conducted with achieving multiple interests. The RECC assesses the ability of an ecosystem to resist external disturbances and maintain its original ecosystem services. The RECC considers multiple factors, including human activities, climate change, and energy structure and consumption. Except for the Bayan mine in inner Mongolia, the support index (evaluation of policies, inputs, and technologies to mitigate environmental impact) was greater than the pressure index (omission of policies, inputs, and technologies to mitigate environmental impact), implying limited environmental impact. The ratio of financial investment in pollution control was an important factor limiting environmental sustainability.

Wang and Liang [33] observed geochemical rare earth element fractionation involving the light and heavy rare earth elements in tailings of the Baotou mine in China. Using the NASC and PAAS values for normalization, the light rare earth elements showed greater normalized patterns. A map of the rare earth element PAAS normalized concentrations exhibited reduced concentrations at an increasing distance from the mine site. The total rare earth element concentrations (Σ REE) of surface samples varied from 156 to 57,000 mg kg^{-1}, with a mean of 4700 mg kg^{-1}. The enrichment factors for all of the rare earth elements displayed values suggesting very high to extremely high enrichment in the east, southeast, and south directions from the mine site, whereas enrichment factors in the northwest direction were indicative of only significant impact. In Baotou, China, Zhou et al. [34] investigated rare earth element concentrations in dust samples and subsequently expressed data using enrichment factors. The enrichment factors, when normalized to the local loess, indicated rare earth element contamination; that is, the igeo index indicated contamination and the HQ did not indicate contamination. In India, Humsa and Srivastava [35] investigated industrial waste from a titanium dioxide pigment industry. Soil resources in the impacted area demonstrated increased heavy metal (Fe, Cr, V, Ni, Cu, Zn, and Pb) concentrations, as well as increased concentrations of Sm, Tb, and Dy.

In Australia, Nkrumah et al. [36] investigated the rare earth element biogeochemical behavior in natural ecosystems to estimate their soil abundances. In addition to a slight HREE enrichment, key rare earth element concentrations were (i) Ce (2550 mg kg^{-1}), (ii) La (645 mg kg^{-1}), (iii) Gd (25 mg kg^{-1}), and Lu (1.5 mg kg^{-1}). Plant uptake showed variation among species, with broadleaf plants typically having greater rare earth element accumulation than greases. In Brazil, Cunha et al. [37] observed uranium-phosphate deposits and estimated the spatial rare earth element soil distribution. The soil rare earth element concentrations were closely correlated with the soil uranium and phosphate concentrations. In China, Zhao et al. [38] evaluated mine tailings from sites with and without phytoremediation. The Nd and Y hazard quotients (HQ) were determined, with the concentration baselines for Nd and Y established at which 10% soil root length inhibition was observed. The HQs for wheat (*Triticum aestivum*) and mung bean (*Vigna radiata*) were less than 1; however, the geo-accumulation index was variably distributed as (i) uncontaminated (Igeo <0), (ii) uncontaminated to moderately contaminated ($0 <$ Igeo <1), (iii) moderately contaminated ($1 <$ Igeo <2), and (iv) moderately to strongly contaminated ($2 <$ Igeo <3). Wang and Liang [39] assessed the environmental impact of rare earth elements in soils surrounding the Bayan Obo Mine. The Bayan Obo deposit is estimated to contain more than 100 million Mt. of rare earth reserves containing monazite and bastnaesite ores. The sum of the rare earth element concentrations in surface soil horizons varied from 150 to 18,900 mg kg^{-1}. The distribution patterns of the individual rare earth elements were like those of the Bayan Obo ores.

MacDonald et al. [40] reviewed the development of pollution indices for freshwater ecosystems, noting that further research is warranted to guarantee accurate predictive environmental outcomes. Chamber [41] discussed "technologically enhanced naturally occurring radioactive material," noting that monazite mining will produce waste material having ^{232}Th. ^{232}Th will alpha decay slowly to ^{228}Ra, which will more rapidly decay by beta emission to ^{228}Ac (half-life of 5.75 years). Where the activity is substantial, appropriate actions to protect the environment and personnel are warranted. Aide and Aide [42] discussed the use of rare earth elements in identifying and assessing soil lithologic discontinuities. Aide and Aide also demonstrated that Fe-Mn masses (pedogenic nodules) in selected alluvial soils preferentially accumulated Ce and revealed a positive Ce anomaly, suggesting that alternating conditions of oxidation–reduction were important for glaebule (Fe-Mn nodules) synthesis and Ce incorporation.

6. Observation of rare earth element soil distributions in missouri

6.1 Study area

Summers are hot and humid with a mean July temperature of 26°C and winter temperatures are mild with a mean January temperature of 2°C. The mean annual precipitation of 1.19 m is seasonally distributed, with greater rainfall in spring. Vegetation is either a mixed hardwood forest or land that has been cleared of vegetation, land-graded, and employed in row-crop agriculture.

6.2 Soils

The alred series (loamy-skeletal over clayey, siliceous, semiactive, mesic Typic Paleudalfs) consists of very deep, well-drained soils formed in cherty hillslope sediments and the underlying clayey residuum. The Alred has an A-E-Bt-2Bt horizon sequence. The rueter series (loamy-skeletal, siliceous, active, mesic Typic Paleudalfs) consists of very deep, somewhat excessively drained soils formed in colluvium and residuum from cherty limestone. The Rueter has an A-E-Bt-2Bt horizon sequence. The Menfro series (Fine-silty, mixed, superactive, and mesic Typic Hapludalfs) consists of very deep, well-drained soils formed in thick loess deposits on uplands. The Menfro has an A-E-BE-Bt-C horizon sequence. The Kaintuck soil series in Missouri (coarse-loamy, siliceous, superactive, nonacid, mesic Typic Udifluvents) are very deep and well-drained floodplain soils formed from loamy alluvium and have an Ap-C horizon sequence.

6.3 Protocols

An aqua regia digestion was employed to obtain a near-total estimation of elemental abundance associated with all but the most recalcitrant soil chemical environments. Aqua regia does not appreciably degrade quartz, albite, orthoclase, anatase, barite, monazite, sphene, chromite, ilmenite, rutile, and cassiterite; however, anorthite and phyllosilicates are partially digested. Homogenized samples (0.75 g) were equilibrated with 0.01 liter of aqua regia (3 mole nitric acid: 1 mole hydrochloric acid) in a 35°C incubator for 24 hours. Samples were shaken, centrifuged, and filtered (0.45 μm), with a known aliquot volume analyzed using inductively coupled plasma mass spectrometry (ICP-MS). Aide and Fasnacht [43] reviewed the application of the aqua regia digestion protocol.

7. Missouri rare earth element investigations

The Alred and Rueter soil profiles were completely characterized, with the rare earth element suite determined with aqua regia digestion. The ochric (A and E horizons) and illuvial (Bt and 2Bt) soil horizons were identified and compared to their rare earth element concentrations. For clarity, eluvial soil horizons occur where strong leaching of clay minerals, oxides, and organic material is observed, whereas illuvial soil horizons contain materials that have been transported downwards either in solution or suspension and subsequently deposited. The illuvial horizons, with their greater clay contents, show greater rare earth element abundances (**Table 4**). Paired t-test comparing ochric and illuvial rare earth element concentrations are significant (0.016 for Alred and 0.020 for Rueter).

The multiple pedons for the Menfro and Kaintuck soil series were characterized for rare earth element concentrations for each soil horizon in their respective soil profiles (**Table 5**). The Menfro soil is an Alfisol with ochric silt loam texture transitioning to a silty clay loam texture in the agrillic horizon. The Kaintuck soil series is an entisol

	Alred		Rueter	
Element	**Ochric**	**Illuvial**	**Ochric**	**Illuvial**
		mg kg^{-1}		
La	16.9	34.6	1.2	6.2
Ce	41.9	85.2	10.1	29.7
Pr	3.7	8.3	0.5	2.0
Nd	13.7	33.3	2.1	9.1
Sm	2.4	6.5	0.4	2.3
Eu	0.3	1.3	0.1	0.5
Gd	2.0	6.4	0.3	2.4
Tb	0.2	0.9	0.1	0.4
Dy	1.2	5.7	0.2	2.5
Ho	0.2	1.2	0.1	0.5
Er	0.5	3.3	0.2	1.5
Tm	0	0.5	0	0.2
Yb	0.4	3.0	0.2	1.5
Lu	0	0.4	0	0.2

Table 4.
The rare earth element distribution of the Alred and Rueter series.

	Menfro series		Kaintuck series	
Element	**Mean (mg kg^{-1})**	**CV (%)**	**Mean (mg kg^{-1})**	**CV (%)**
La	25.8	13.6	17.2	6.0
Ce	43.9	12.6	43.7	3.1
Pr	6.2	15.5	4.6	4.9
Nd	23.2	17.4	19.4	4.9
Sm	4.4	20.3	3.2	9.0
Eu	0.8	23.7	0.6	<1
Gd	3.5	22.5	2.9	5.3
Tb	0.4	26.4	0.4	14.4
Dy	2.1	27.0	2.3	6.5
Ho	0.4	27.2	0.4	<1
Er	0.9	30.9	1.2	7.1
Tm	0.1	28.3	0.1	<1
Yb	0.8	32.0	0.9	5.6
Lu	0.1	136	0.1	<1

Table 5.
The mean rare earth element concentrations and coefficient of variations.

with sandy loam soil textures througthout the soil profiles. The Menfro soil profiles have greater rare earth element concentrations, a feature attributed to their greater clay contents. The coefficient of variation for the rare earth element concentrations is greater for the Menfro soil pedon, a feature attributed to the discrete clay differences in the ochric and agrillic horisons when compared to the rather uniform texture distribution in the Kaintuck pedons.

The Wilbur soil series (coarse-silty, mixed, superactive, mesic Fluvaquentic Eutrudepts) consists of very deep, moderately well-drained soils formed in silty alluvium. The four pedons are located across a 40 ha production field and were evaluated for morphology, routine physical and chemical characteristics, and rare earth element concentrations across all pedon horizons. Yttrium was included in the rare earth element analysis. As expected, the rare earth element distribution clearly is in accordance with the Oddon-Harkin rule (**Table 6**).

The Wilbur pedons are distant from industry and mining, thus these pedons may be considered pristine. The rare earth element distribution is very uniform within and across pedons, features reflecting the pedon's as belonging to the Inceptisol order and having little soil texture variation. The rare earth elements may be used in the GeoBase for the following pollution indices: (i) geoaccumulation index, (ii) single pollution index, (iii) The enrichment factor, (iv) the contamination factor, (v) the sum of contamination, (vi) the pollution load index, (vii) the average single pollution index, (viii) the vector modulus pollution index, (ix) the degree of contamination, and (x) the hazard quotient. If an estimate of the factor of the rare earth element's toxic reaction becomes available, then (i) the potential ecological risk factor and (ii) the potential ecological hazard indexes can be calculated.

Element	Mean	Coefficient variation
	mg kg^{-1}	Percent
Y	12.1	21
La	27.3	7
Ce	57.3	7
Pr	6.9	8
Nd	25.9	9
Sm	4.7	10
Eu	0.9	15
Gd	4.1	14
Tb	0.5	18
Dy	2.7	17
Ho	0.5	21
Er	1.3	19
Tm	0.2	28
Yb	1.0	19
Lu	0.1	34

Total of 32 observations of four pedons.

Table 6.
Mean and coefficient variation for the Wilbur soil series.

8. Prospectus concerning research needs

New rare earth element mines are in various states of development across the United States and in many other nations. Given that rare earth element mines and ore processing operations span a range of locations, the climate, and physical settings will vary. The environmental impact assessment must include (i) mine conditions, (ii) ore geochemistry, (iii) local geology and hydrology, and (iv) physiographic settings and geomorphology. Mine and processing sites may disturb or accentuate (i) surface water because of drilling fluids, acid, and neutral mine discharge water, and influence aquatic organisms, (ii) groundwater impacts because of mine pit lakes, evaporation ponds, (iii) air pathways involving fugitive dust, aerosols and chemical vapors, radioactivity, and (iv) tailing storage facilities [44]. The environmental assessment must consider (i) current and future construction workers/employees (ingestion and inhalation), (ii) traditional tribal lifeways, and (iii) on-site and off-site residents. The documentation of potential health effects requires a greater research emphasis. Pulmonary toxicity of inhaled rare earth elements, and expanded investigations, involving integrated risk information systems and provisional peer-reviewed toxicity values are necessary [44].

Pollution indices need to be verified for the rare earth elements. Additionally, substantial research in assessing the biological availability of the rare earth elements in ecosystems is warranted as the total rare earth element concentrations may not accurately describe the environmental risk.

Author details

Michael T. Aide
Southeast Missouri State University, Cape Girardeau, USA

*Address all correspondence to: mtaide@semo.edu

References

[1] Lee JD. Concise Inorganic Chemistry. NY: Chapman and Hall; 1992

[2] Aide MT. Lanthanide soil chemistry and its importance in understanding soil pathways: Mobility, plant uptake and soil health. In: Lanthanides. London, UK, Rijeka, Croatia: InTech; 2018

[3] Kabata-Pendias A. Trace Elements in Soils and Plants. Boca Raton, FL: CRC Press; 2011

[4] McLennan SM. Rare earth elements in sedimentary rocks: Influence of provenance and sedimentary processes. In: Lipin BR, McKay GA, editors. Geochemistry and mineralogy of rare earth elements. Reviews in Mineralogy, Mineral. Vol. 21. Washington, DC: Soc. Am.; 1989

[5] McLennan SM. Relationship between the trace element composition of sedimentary rocks and upper continental crust. Geochemistry, Geophysics, and Geosystems. 2001;**2**. Article No. 2000GC00109. DOI: 10.1029/2000GC000109

[6] Rudnick RL, Gao S. Composition of the continental crust, treatise on geochemistry. Treatise on Geochemistry. 2003;**3**:1-64. DOI: 10.1016/B0-08-043751-6/03016-4

[7] Van Gosen BS, Verplanck PL, Seal RR II, Long KR, Gambogi J. Rare-earth elements, chap. O. In: Schulz KJ, DeYoung JH Jr, Seal RR II, Bradley DC, editors. Critical mineral resources of the United States—Economic and Environmental Geology and Prospects for Future Supply: U.S. Geological Survey Professional Paper 1802. Washington DC: United States Geological Survey; 2017. pp. O1-O31. DOI: 10.3133/pp1802

[8] U.S. Geological Survey Mineral commodity summaries. January 2022. Available from: https://pubs.usgs.gov/periodicals/mcs2022/mcs2022-rare-earths.pdf

[9] Ramos SJ, Dinali GS, Oliveira C, Martins GC, Mureira CG, Siqueira JO, et al. Rare earth elements in the soil environment. Current Pollution Reports. 2016;**2**:28-50. DOI: 10.1007/s40726-016-0026-4

[10] Kim H-M, Jariwala D. The Not so Rare Earth Elements: A Question of Supply and Demand. Philadelphia, PA: Kleinman Center for Energy Policy. University Pennsylvania; 2021. Available from: https://kleinmanenergy.upenn.edu/wp-content/uploads/2021/09/KCEP-The-Not-So-Rare-Earth-Elements_withES.pdf

[11] Van Veen K, Melton A. Rare Earth elements supply chains, Part 1: An update on global production and trade. United States International Trade Commission Executive Briefing on Trade, 2020. 2020. Available from: https://www.usitc.gov/publications/332/executive_briefings/ebot_rare_earths_part_1.pdf

[12] Rim K. Effects of rare earth elements on the environment and human health: A literature review. Toxicology and Environmental Health Sciences. 2016;**8**:189-200. DOI: 10.1007/s13530-016-0276-y

[13] Pagano G, Thomas PJ, Di Nunzio A, Trifuoggi M. Human exposures to rare earth elements: Present knowledge and research prospects. Environmental Research. 2019;**171**:493-500. DOI: 10.1016/j.envres.2019.02.004

[14] Landim JSP, da Silva YJAB, do Nascimento CWA, da Silva YJAB,

Nascimento RC, Boechat CL, et al. Distribution of rare earth elements in soils of contrasting geological and pedological settings to support human health assessment and environmental policies. Environmental Geochemistry and Health. 2022;**44**(3):861-872. DOI: 10.1007/s10653-021-00993-0

[15] González V, Vignati DA, Pons MN, Montarges-Pelletier E, Bojic C, Giamberini L. Lanthanide ecotoxicity: First attempt to measure environmental risk for aquatic organisms. Environmental Pollution. 2015;**199**:139-147. DOI: 10.1016/j.envpol.2015.01.020

[16] Malhotra N, Hsu HS, Liang ST, Roldan MJM, Lee JS, Ger TR, et al. An updated review of toxicity effect of the rare earth elements (REEs) on aquatic organisms. Animals (Basel). 2020;**10**(9):1663. DOI: 10.3390/ani10091663

[17] Thomas PJ, Carpenter D, Boutin C, Allison JE. Rare earth elements (REEs): Effects on germination and growth of selected crop and native plant species. Chemosphere. 2014;**96**:57-66. DOI: 10.1016/j.chemosphere.2013.07.020

[18] Li J, Verweij RA, van Gestel CAM. Lanthanum toxicity to five different species of soil invertebrates in relation to availability in soil. Chemosphere. 2018;**193**:412-420. DOI: 10.1016/j.chemosphere.2017.11.040

[19] Zhou L, Li Z, Liu W, Liu S, Zhang L, Zhong L, et al. Restoration of rare earth mine areas: Organic amendments and phytoremediation. Environmental Science and Pollution Research International. 2015;**22**(21):17151-17160. DOI: 10.1007/s11356-015-4875-y

[20] Askari MS, Alamdari P, Chahardoli S, Afshari A. Quantification of heavy metal pollution for environmental assessment of soil condition. Environmental Monitoring and Assessment. 2020;**192**:1-17. DOI: 10.1007/s10661-020-8116-6

[21] Kowalska JB, Mazurek R, Gąsiorek M, Zaleski T. Pollution indices as useful tools for the comprehensive evaluation of the degree of soil contamination–A review. Environmental Geochemistry and Health. 2018;**40**:2395-2420. DOI: 10.1007/s10653-018-0106-z

[22] Santos-Francés F, Martinez-Graña A, Alonso Rojo P, García SA. Geochemical background and baseline values determination and spatial distribution of heavy metal pollution in soils of the andes mountain range (Cajamarca-Huancavelica, Peru). International Journal of Environmental Research and Public Health. 2017;**14**(8):859. 10.3390/ijerph14080859

[23] Jiang HH, Cai LM, Wen HH, Luo J. Characterizing pollution and source identification of heavy metals in soils using geochemical baseline and PMF approach. Scientific Reports. 2020;**10**:6460. DOI: 10.1038/s41598-020-63604-5

[24] Diwa RR, Elvira MV, Deocaris CC, Fukuyama M, Belo LP. Transport of toxic metals in the bottom sediments and health risk assessment of Corbicula fluminea (Asiatic clam) collected from Laguna de Bay, Philippines. Science of The Total Environment. 2020;**838**(Pt 4):156522. DOI: 10.1016/j.scitotenv.2022.156522

[25] Elvira MV, Faustino-Eslava DV, Fukuyama M, de Chavez EC, Padrones JT. Ecological risk assessment of heavy metals in the bottom sediments of Laguna de Bay, Philippines. The Mindanao Journal of Science and Technology. 2020;**18**(2):311-335

[26] Lawrence V, Tannenbaum M, Johnson S, Bazar M. Application of the

hazard quotient method in remedial decisions: A comparison of human and ecological risk assessments. Human and Ecological Risk Assessment: An International Journal. 2003;**9**(1):387-401. DOI: 10.1080/713609871

[27] Barbieri M. The importance of enrichment factor (EF) and geoaccumulation index (Igeo) to evaluate the soil contamination. Journal of Geology and Geophysics. 2016;**5**:1. DOI: 10.4172/2381-8719.1000237

[28] Ghrefat HA, Abu-Rukah Y, Rosen MA. Application of geoaccumulation index and enrichment factor for assessing metal contamination in the sediments of Kafrain Dam, Jordan. Environmental Monitoring and Assessment. 2011;**178**:95-109. DOI: 10.1007/s10661-010-1675-1

[29] Gargouri D, Gzam M, Kharroubi A, Jedoui Y. Use of sediment quality indicators for heavy metals contamination and ecological risk assessment in urbanized coastal zones. Environment and Earth Science. 2018;**77**:381. DOI: 10.1007/s12665-018-7567-3

[30] Weissmannová HD, Mihočová S, Chovanec P, Pavlovský J. Potential ecological risk and human health risk assessment of heavy metal pollution in industrial affected soils by coal mining and metallurgy in Ostrava, Czech Republic. International Journal of Environmental Research and Public Health. 2019;**16**(22):4495. DOI: 10.3390/ijerph16224495

[31] Krasavtseva E, Maksimova V, Makarov D. Conditions affecting the release of heavy and rare earth metals from the mine tailings Kola subarctic. Toxics. 2021;**9**:163. DOI: 10.3390/toxics9070163

[32] Bai J, Xu X, Duan Y, Zhang G, Wang Z, Wang L, et al. Evaluation of resource and environmental carrying capacity in rare earth mining areas in China. Scientific Reports. 2022, 2022;**12**:6105. DOI: 10.1038/s41598-022-10105-2

[33] Wang L, Liang T. Geochemical fractions of rare earth elements in soil around a mine tailing in Baotou, China. Scientific Reports. 2015;**5**:12483. DOI: 10.1038/srep12483

[34] Zhou H, Chun X, Lu C, He J, Du D. Geochemical characteristics of rare earth elements in windowsill dust in Baotou, China: Influence of the smelting industry on levels and composition. Environmental Science: Processes & Impacts. 2020;**22**:2398. DOI: 10.1039/d0em00273a

[35] Humsa TZ, Srivastava RK. Impact of rare earth mining and processing on soil and water environment at Chavara, Kollam, Kerala: A case study. Procedia Earth and Planetary Science. 2015;**11**:566-581. DOI: 10.1016/j.proeps.2015.06.059

[36] Nkrumah PN, Erskine PD, Erskine JD, van der Ent A. Rare earth elements (REE) in soils and plants of a uranium-REE mine site and exploration target in Central Queensland, Australia. Plant and Soil. 2021;**464**:375-389. DOI: 10.1007/s11104-021-04956-3

[37] Cunha CSM, da Silva YJAB, Escobar MEO, do Nascimento CWA. Spatial variability and geochemistry of rare earth elements in soils from the largest uranium-phosphate deposit of Brazil. Environmental Geochemistry and Health. 2018;**40**(4):1629-1643. DOI: 10.1007/s10653-018-0077-0

[38] Zhao CM, Shi X, Xie SQ, Liu WS, He EK, Tang YT, et al. Ecological

risk assessment of neodymium and yttrium on rare earth element mine sites in Ganzhou, China. Bulletin of Environmental Contamination and Toxicology. 2019;**103**:565-570. DOI: 10.1007/s00128-019-02690-2

[39] Wang L, Liang T. Anomalous abundance and redistribution patterns of rare earth elements in soils of a mining area in Inner Mongolia, China. Environmental Science and Pollution Research International. 2016;**23**(11):11330-11338. DOI: 10.1007/s11356-016-6351-8

[40] MacDonald DD, Ingersoll CG, Berger TA. Development and evaluation of consensus-based sediment quality guidelines for freshwater ecosystems. Archives of Environmental Contamination and Toxicology. 2000;**39**:20-31. DOI: 10.1007/s002440010075

[41] Chambers DB. Radiological protection in North American naturally occurring radioactive material industries. Annal of the ICRP. 2015;**44**(1_suppl). DOI: 10.1177/0146645315572300

[42] Aide MT, Aide CC. Rare earth elements: Their importance in understanding soil genesis. International Scholarly Research Notices Network (ISRN Soil Science). 2012;**2012**. Article ID783876. DOI: 10.5402/2012/783876

[43] Aide MT, Fasnacht M. Estimating trace element availability in soils having a seasonal water table using commercially available protocols. Communications in Soil Science and Plant Analysis. 2010;**41**:1159-1177. DOI: 10:1080/00103621003721379

[44] Reisman D, Weber R, McKernan J, Northeim C. Rare earth Elements: A Review of Production, Processing, Recycling, and Associated Environmental Issues. Washington DC. EPA 600/R-12/572: United States Environmental Protection Agency; 2012. Available from: www.epa.gov/ord